AF696811

ESSAYS OF A SOVIET SCIENTIST

Masters of Modern Physics

Advisory Board

Dale Corson, Cornell University
Samuel Devons, Columbia University
Sidney Drell, Stanford Linear Accelerator Center
Herman Feshbach, Massachusetts Institute of Technology
Marvin Goldberger, University of California, Los Angeles
Wolfgang Panofsky, Stanford Linear Accelerator Center
William Press, Harvard University

Series Editor

Robert N. Ubell

Published Volumes

The Road from Los Alamos by Hans A. Bethe
The Charm of Physics by Sheldon L. Glashow
Citizen Scientist by Frank von Hippel
Visit to a Small Universe by Virginia Trimble
Nuclear Reactions: Science and Trans-Science by Alvin M. Weinberg
In the Shadow of the Bomb: Physics and Arms Control by Sydney D. Drell
The Eye of Heaven: Ptolemy, Copernicus, and Kepler by Owen Gingerich
Particles and Policy by Wolfgang K. H. Panofsky
At Home in the Universe by John A. Wheeler
Cosmic Enigmas by Joseph Silk
Nothing Is Too Wonderful To Be True by Philip Morrison
Arms and the Physicist by Herbert F. York
Confessions of a Technophile by Lewis M. Branscomb
Making Waves by Charles H. Townes
Of One Mind: The Collectivization of Science by John Ziman
Einstein, History, and Other Passions by Gerald Holton
Bright Galaxies, Dark Matters by Vera Rubin
Atomic Histories by Rudolf E. Peierls
Essays of a Soviet Scientist by Vitaliĭ I. Gol'danskiĭ

ESSAYS OF A SOVIET SCIENTIST

Vitaliĭ I. Gol'danskiĭ

N. N. Semenov Institute of Chemical Physics
of the Russian Academy of Sciences
Moscow, Russia

AIP PRESS

American Institute of Physics Woodbury, New York

In recognition of the importance of preserving what has been written, it is a policy of the American Institute of Physics to have books published in the United States printed on acid-free paper.

©1997 by American Institute of Physics
All rights reserved.
Printed in the United States of America.

Reproduction or translation of any part of this work beyond that permitted by Section 107 or 108 of the 1976 United States Copyright act without the permission of the copyright owner is unlawful. Requests for permission or further information should be addressed to the Office of Rights and Permissions, 500 Sunnyside Boulevard, Woodbury, NY 11797-2999; phone: 516-576-2268; fax: 516-576-2499; e-mail: rights@aip.org.

AIP Press
American Institute of Physics
500 Sunnyside Boulevard
Woodbury, NY 11797-2999

Library of Congress Cataloging-in-Publication Data
Gol'danskiĭ, V. I. (Vitaliĭ Iosifovich)
Essays of a Soviet Scientist / Vitaliĭ I. Gol'danskiĭ.
p. cm. -- (Masters of Modern Physics)
Includes index.
ISBN 1-56396-454-6
1. Science--Soviet Union--History--20th century. 2. Science-Social aspects--Soviet Union. 3. Science and state--Soviet Union. 4. Scientists--Soviet Union. I. Title. II. Series.
Q127.S696G65 1996 96-17869
509.47--dc20 CIP

10 9 8 7 6 5 4 3 2 1

This book is volume nineteen of the Masters of Modern Physics Series.

Contents

Art, Music, Literature, and the Media

PART II. *SPEAKING OUT*

For Peace against Nuclear Threat

Politics

About the Series

Masters of Modern Physics introduces the work and thought of some of the most celebrated physicists of our day. These collected essays offer a panoramic tour of the way science works, how it affects our lives, and what it means to those who practice it. Authors report from the horizons of modern research, provide engaging sketches of friends and colleagues, and reflect on the social, economic, and political consequences of the scientific and technical enterprise.

Authors have been selected for their contributions to science and for their keen ability to communicate to the general reader—often with wit, frequently in fine literary style. All have been honored by their peers and most have been prominent in shaping debates in science, technology, and public policy. Some have achieved distinction in social and cultural spheres outside the laboratory.

Many essays are drawn from popular and scientific magazines, newspapers, and journals. Still others—written for the series or drawn from notes for other occasions—appear for the first time. Authors have provided introductions and, where appropriate, annotations. Once selected for inclusion, the essays are carefully edited and updated so that each volume emerges as a finely shaped work.

Masters of Modern Physics is edited by Robert N. Ubell and overseen by an advisory panel of distinguished physicists. Sponsored by the American Institute of Physics, a consortium of major physics societies, the series serves as an authoritative survey of the people and ideas that have shaped twentieth-century science and society.

Preface

So many shocks, so many revolutions in mentality, in definitions of good and evil, have fallen to my generation's lot in the USSR, in Russia, that sometimes I can hardly perceive myself in various periods of life as the same "myself."

I was born in 1923 into the family of a Jewish teacher. We lived in the town of Vitebsk, now famous everywhere thanks to Marc Chagall. My parents, Iosif and Judith', and my elder sister Mira are always alive in my grateful memory. Though non-Party, we were, up to a certain time, true believers in the ideals of communism—in Stalin's grandeur and infallibility. Their faith was conveyed to me too, and parting with that faith was a long and excruciating process.

Elementary school and my first two undergraduate years were spent at the Chemistry Faculty of the university of Leningrad. The years 1937–38 turned out to be very painful for our family—two brothers and a brother-in-law of my mother were shot and, being unable to cope with this loss, my father died of a heart attack.

The faith my parents and sister had in Stalin was irrevocably broken, but in trying to protect me from shock they did not share their grievous thoughts with me. As teenagers, at that time, our passionate dream was to find ourselves in Madrid, to fight against fascists for our romantic ideal—a Spanish republic.

Following a long-standing Russian tradition, I continued to believe that our "Tsar is good and the ministers are bad."

I even tried to find some explanations for myself, some justifications for the loathsome Molotov-Ribbentrop Pact, and the nazist invasion on June 22, 1941, brought back my reckless faith in Stalin as the sole power able to smash the hateful devil—Hitler.

Reminiscences of years spent during the war seem unreal today, as if they happened in another dimension: being wounded by a Luftwaffe bullet while digging trenches near Novgorod, on the brink of death by starvation during the winter of 1941–42 at the siege of Leningrad (it took away several of our relatives), and the evacuation by the ice "life-road" across the Ladoga lake, a three-week long journey in a frozen cargo-wagon to Kazan' at Volga.

In April 1942 I resumed my undergraduate studies at the Chemistry Faculty of the local university and also began working for the Academy of Sciences. Starting as a humble laboratory assistant there, my career progressed over the years until I became the Director of the Institute of Chemical Physics and Academician.

Meanwhile our country was, as usual, rocked between scorching heat and icy frost, between timid hopes and deep despair; great victory in war and, generated by victory, great expectations—hopes for the new way of life, for real democracy ... Bitter disappointments—the wrecking of the creative intelligentsia, the wrecking of biology, the rupture with war allies, and the iron curtain—the enslavement of Eastern Europe, the new wave of terror, and the flourishing and encouragement of State anti-semitism ... However, I should confess—to my shame—that even at that time I still believed that the aging Stalin was the last obstacle standing in the way of younger black-hundred pogrom activists longing for full power and I perceived Stalin's death as misfortune.

Sight was regained only when Khrushchev unmasked Stalin as an arrant criminal and butcher. I count myself among the so-called "people of the sixties," i.e., among those who enthusiastically met "Khrushchev's thaw" as a first step towards the "socialism with human face." However, the pendulum of our history continued its swing.

The bitter and disgraceful day came when our tanks crushed "Prague Spring." Eighteen years of rotten Brezhnev's regime generated feelings of hopelessness and apathy, relay-race deaths of three party leaders and, finally, the appearance of Gorbachev at the proscenium. The euphoria of perestroika and glasnost. For 44 months—until the last day of the existence of the USSR—I was a member of the first and last Congress of People's Deputies of the Soviet Union and for the first time literally plunged my soul into politics—defending our young democracy at the Parliament's floor, on Television, and in various mass media, arguing for freedom, for the nuclear-weapon-free world, and against militarism.

After the break-up of the Soviet Union in December 1991 I stepped out of "great politics." I still meet with Gorbachev on friendly terms from time to time, but I do not advise him to aspire to return to the corridors of power. Whatever the general prospects for the future of Russia may be, the role of Cassandra does not tempt me.

I feel a certain resemblance between my stories in life and in science. In both cases my aspirations have not always been the same. My earliest passion was history, at about 10–15 years of age. From time to time, new passions appeared, new infatuations which flung me from one problem to another and brought me finally to a peculiar "delocalization" between chemistry, physics, and biology.

In 1944 I graduated from the Chemistry Faculty of the Moscow University. My diploma work as well as my Ph.D. (candidate of sciences) thesis were devoted to purely chemical problems—to catalysis.

My supervisor during my postgraduate studies at the Institute of Chemical Physics (ICP) and generally my main guide was our brilliant scientist—I even take the liberty of calling him a genius—Nikolaĭ Nikolaevich Semenov. He received the Nobel Prize in Chemistry in 1956.

After Hiroshima the ICP, as many other institutes, transferred its attention to nuclear problems, and I was charged with the task of dealing with neutron physics. Our 1949–50 experiments at the Dubna synchrocyclotron have shown that the bombardment of large uranium targets by particles of hundreds of MeV energy leads to the formation of several scores of secondary neutrons (i.e., of plutonium nuclei) for each projectile. These (and simultaneous independent American) experiments later laid the grounds for projects involving electrical generation of nuclear fuel and new ways for producing atomic energy.

Several years of my life during the 1950s were spent in the P. N. Lebedev Physical Institute (FIAN), which became an invaluable school for me in many fields of physics.

In 1954 my first publication appeared in our best physical journal, JETP. Its subject was completely different from our Dubna experiments (at that time still classified)—Čerenkov radiation of cosmic particles in atmosphere and the idea of a new method of studies of extensive air showers.

My experimental group at the FIAN synchrotron has enthusiastically studied elastic γP-scattering, Compton effect on protons. In 1959 we published our results—the main result was the determination of the numerical value of the electric polarizability of protons (still not revised).

Meanwhile I became interested in a variety of subjects, new passions which literally jostled each other.

Based on the principle of isotopic invariance of nuclear forces I have obtained the formula for masses of neutrondeficient nuclei (which anticipated the well-known Garvey-Kelson formula) and predicted and described some new nuclear phenomena—two-proton radioactivity, emission of delayed proton and neutron pairs, and the nuclear Josephson effect.

I was delighted by the discovery of the Mössbauer effect and conceived a desire to apply it to a broad range of chemical problems. Being educated by the Semenov school in electron chemistry and in the reactions of free atoms and radicals, I aspired also to the chemical applications of the annihilation of positrons. Thus, at the brink of the 1960s I entered two of the future directions of my investigations at the border between physics and chemistry—Mössbauer chemical spectroscopy (Monographs, 1963 and 1968) and physical chemistry of positrons and positronium (Monograph, 1968).

I felt drawn back to chemistry now, with the use of knowledge accumulated in physics, and in 1959 resumed experimental works in my scientific *Alma mater*, the Institute of Chemical Physics.

In connection with the studies of solid-phase, low-temperature reactions of polymerization I took a vivid interest in the role of the quantum tunnel effect in chemical conversions. At that time this effect was studied exclusively in liquid phase by the kinetical *H/D* isotope effect. In two 1959 papers I obtained the expression for the so-called "cross-over temperature" T_c (at $T < T_c$ the tunneling dominates classical overbarrier transitions) and came to the conclusion on the existence of a finite low-temperature limit of the chemical reaction rate. Such a limit was experimentally discovered in our laboratory (1973) by the example of radiation-induced polymerization of formaldehyde and explained as molecular tunneling. On the base of molecular tunneling I put forward the hypothesis of "cold prehistory of life"—the revival of extraterrestrial, panspermia scenario of the origin of life (Svante Arrhenius, 1908).

Experimental and theoretical studies of tunneling phenomena in chemical physics (including the 1986 Monograph, and later treatment of multidimensional tunneling) turned out to be the leading direction of my scientific interests for decades, and generally of my contributions to chemical physics.

Research in Mössbauer chemical spectroscopy, where we have enjoyed close collaboration with Rudolf Mössbauer and his school since the end of the 1960s, brought us to the studies of the dynamics of biopolymers—to the glass-like model of their dynamical properties which obtained wide recognition.

On the other hand, the hypothesis of the cold prehistory of life and our earlier experiments (1962) on the irradiation of *L*- and *D*-diastereomers by β^--particles were to some extent an introduction to the field which now apparently became my main (and, at 72 years, I also think my last) hobby—chirality, the origin of life, evolution.

Here, I believe, is a proper place to end this preface.

However, before I do, I want to express my cordial thanks to my teachers, my friends, my collaborators, my coauthors, my disciples, and my wife Lioudmila (we hope to celebrate our golden wedding anniversary in 1997). Without all of them my life would have been an empty, shapeless envelope, deprived of any content.

Our century is approaching its end. It was the most horrible century in the history of humankind, surpassing by far the totality of all preceding centuries by the amount of violent deaths. At the same time it was the century of an unprecedented flourishing of science, of the greatest scientific and technical accomplishments. In the future, will the top achievements of the human mind be helpless against the abysses of human madness? I want to believe that we are not helpless, so that we can boldly enter the 21st Century, the Third Millenium, with the hopes for a really bright future for humankind.

PART I. ESSAYS

REMINISCENCES

The main fraction of this section is composed of my contributions to the books published by our Academy of Sciences (Publishing House "Nauka") in the series "Reminiscences on Scientists." These books are dedicated to the memory of our most distinguished academicians and are usually compiled by the closest relatives, friends, or disciples of the deceased who invite contributors of their choice.

A Living Legend
(Lev Andrejevich Artsimovich)

On April 10, 1973, 40 days after the death of Lev Artisimovich, several dozen people gathered to honor his memory in the apartment, well known to all, with the cast-iron lion (*lev,* in Russian) at the door (a gift from Anatoli Petrovich Aleksandrov for Lev Andreevich's 60th birthday).

Each person said only a few words at the mourning table, but altogether these words produced a vivid image of the unique, vibrant personality of Artsimovich and the loss felt by all those present.

What a pity that everything said that evening was never written down. Our memory is so flawed, and today many of those who described their impressions and recollections of Lev then are no longer living. For example, I recall how Rem V. Khokhlov, who had become the rector of Moscow State University, in speaking of the exceptional love the students had for Artsimovich as a lecturer of incomparable depth and clarity, recalled a story circulating among the students: how Lev, already a venerable professor, drank a glass of mercury in lecture, in order to demonstrate more clearly the unusual properties of this metal.

My first memories associated with the name of Artsimovich are also of a legendary figure. However, many years later, after having heard from Artsimovich himself about the time when these "legends" were created. I understood that they are, in fact, very close to the truth.

August, 1941. A trainload of people from the Physico-Technical Institute, evacuated from Leningrad a few days before the closing of the blockade ring, slowly moves toward Kazan. The road is long and oppressive, and the spirits of many are low. Something is needed to cheer everyone up. And, in fact, something unexpected occurs: news spreads of a forthcoming duel between two Leningraders!

A colleague at the Institute of Chemical Physics, S.R., (completely devoid of humor), had accidentally, (in an encounter arranged by the victim!) shoved Boris P. Aleksandrov of the Physico-Technical Institute on the platform at one of the stations. Lev soon visits S.R. in his role as second for his insulted friend. The formal challenge to a duel is made, with a request that a place, time, and type of weapon be chosen. Frightened by the appearance of the perfectly serious second, S.R. rushes to his director—Nikolaĭ N. Semenov—to ask him to intervene to avert bloodshed. Not immediately, but still believing the unprecedented story, Semenov in turn goes to Abram F. Ioffe and requests his intervention. However, Abram Fedorovich, who had been told everything, answers in a dignified manner: “I’m sorry, Nikolaĭ Nikolaevich, but I cannot interfere in affairs of honor.”

The trip to Kazan goes into its second week. During a long stop at a station, Artsimovich and several friends try to take a long-desired bath. However, their plans are thwarted by the bathing attendant: the gate is locked and he refuses to open it, telling them that the bath is closed on Sundays. Then they climb over the fence and set about their business without permission. The furious bathing attendant brings a militiaman. A big row is about to develop. But then L.A. steps forward and, in his characteristic sarcastic voice, pronouncing the *r*’s in the French style (I hear it now, as I write these lines), asks the militiaman if he really does not know of the decree that employees of the USSR Academy of Sciences were allowed to bathe for free on Sundays. The bewildered policeman retreats, and the bathing-laundry campaign is concluded successfully.

My first direct recollections of L.A. are from 1955. At that time, while giving a course of lectures on nuclear reactions at the Moscow Institute for Physical Engineering, I introduced into the course a brief description of thermonuclear reactions: the Atkinson-Houtermans formula, the Bethe cycle, a thermonuclear “hydrogen” explosion.

One day I was urgently called to the telephone in the main office of the Lebedev Institute, where I was then working. Trying to guess why, I rushed to the second floor and recognized the voice of L.A. He asked me to write an article on thermonuclear reactions for the Large Soviet Encyclopedia, in the physics editorial department of which his first wife was working. “Mariya Nikolaevna asked me to find an author, and I recommended you to her, so please don’t let me down.” I was extremely flattered, because I had already found out that just L.A. was directing the research on controlled thermonuclear reactions in our country. But then he explained: “Of course, I or one of my colleagues could write it better, but—you will understand—it is inconvenient.”

Two or three times in the course of writing the article I managed to talk to L.A. about thermonuclear reactions, and got to hear his stories about those classic studies which soon became widely known from the report of Igor' V. Kurchatov at Harwell in England. Only many years later, when L.A. and I became close friends during the last four years of his life, did he speak frankly about the complex mixture of happiness and bitterness, pride and jealousy which he experienced in April 1956, when he himself did not speak but only heard from afar the report on the work of many years of his large group on plasma heating by intense pulses and, essentially as a bystander, witnessed the sensational impression they had made throughout the world. However, his recognition of his indisputable authority in the scientific world, that is to say, his understanding of his own scientific power, allowed L.A. to suppress all such jealousy; he was completely devoid of envy of anyone else's success. I witnessed this once again when, after meeting in the airport in New York in April 1969 for our return to Moscow, the four of us—Lev Andreevich, his wife Nelli Georgievna, Artem I. Alikhan'yan, and I—read in the *New York Times* the news that one of Lev Andreevich's talented students and colleagues, Mikhail S. Ioffe, had won an extremely important international scientific award for his original solution of important problems in controlled thermonuclear fusion, i.e., for only one term in the series of world-class achievements forever associated with the name of Artsimovich.

However, let us now return to the 1950s. The 20th session of the Communist Party of the Soviet Union ushered in a period of rapid development of international scientific relations with the USSR Academy of Sciences. Artsimovich promoted this development very actively as deputy academician-secretary and then as academician-secretary of the Division of Physical and Mathematical Sciences.

In early spring 1957 a group of Soviet scientists, including the present author, was supposed to go to the United States to participate in the Rochester Conference on High Energy Physics.

It should be recalled that during the Cold War the whole subject of nuclear research attracted special interest even in circles far from science, since the scope and level of such research could sometimes give definite information about the nature of applied research. It is therefore not surprising that rather lively arguments erupted over whether or not it was worthwhile for our scientists to go to Rochester. Artsimovich and Kurchatov actively supported going. The trip to Rochester did not take place, but this was only a minor episode in the overall rapid development of international contacts with Soviet science.

In those days I was lucky to meet Artsimovich rather often. His residence at that time was a little house to the right of the main building of the Presidium of the USSR Academy of Sciences. His office and anteroom housing the Division staff were also very small and cozy; there was hardly space for the several telephones on the small stand next to the desk. L.A. dialed one number after another, argued heatedly, and defended his point of view to anyone.

In December 1957, at the reception in the conference hall of the Academy of Sciences on the occasion of the first large post-war international nuclear conference hosted by us in our country, L.A. caustically ridiculed both foreign skeptics who did not believe in the possibility of great scientific achievements in the USSR and were shamed on the historic day that our country launched the first satellite, and the "superpatriots" who had immediately begun to cite the satellite as absolute proof that international scientific contacts were unnecessary. "Why do you need to go to America?" L.A. mimicked such bureaucrats, "After all, it's a wild country, there's nothing there but whiskey and cowboys."

Absorbed in his thermonuclear research, L.A. always retained a lively interest in nuclear physics as such. I remember his outstanding speech at the opening of the International Conference on Nuclear Reactions in the auditorium of Moscow State University in July 1960. At that time one of the main problems in nuclear physics was reactions between complex nuclei, which were studied using beams of accelerated multiply charged ions. He jokingly compared this to studying the inner workings of an automobile from the debris remaining after a collision.

Later in his life L.A. showed great interest in the diverse possibilities of using the techniques and apparatus of nuclear physics for nonnuclear purposes. In October 1972 he, together with me and Serguei P. Kapitsa, wrote a letter to the president of the USSR Academy of Sciences on "Ensuring the experimental development of the natural sciences and the introduction of scientific advances into the national economy on the basis of the Applied Nuclear Science Center of the USSR Academy of Sciences." However, by this time L.A.'s strength was, alas, on the decline, and the proposal did not come to life.

In rereading the above lines, I hesitated at the words "defended his point of view to anyone." These words are very important for describing L.A.'s relations with people.

He was completely free of the very widespread "anisotropic" behavior, i.e., in plain words, his reaction to external irritants did not depend on whether they came from a person occupying a high, intermediate, or low position on the social ladder. In addition, L.A. did not renounce his principles when they came into conflict with personal feelings, even in the case of his closest friends. And, to the credit of his friends, it must be said that they were not offended by this. Truly objective people could not fail to recognize the unvarying and characteristic objectivity of unfavorable judgments passed on them by Artsimovich.

He also attended sessions of the Committee on the Lenin Prize and the State Prize of the USSR in science and technology defending his point of view to anyone. He loved to recall, in particular, how he had blocked the awarding of a prize for work on the construction of the Volgo-Baltic waterway. In his speech he said that this work was certainly distinguished and should be highly valued, but an additional person should be included in the set of authors. He was asked to specify whom he had in mind. L.A. answered immediately: "Pyotr Alekseevich Romanov." "Who is that?" they asked. "Peter the Great. After all, he was the first to begin work on that canal."

A typical manifestation of his objectivity was his skill at recognizing his own mistakes in a scientific controversy or in dealings with people. For example, having become acquainted in the fall of 1972 in Paris (at the house of his friends Edmond and Marguerite Grillot) with our distinguished historian Albert Z. Manfred, and finding himself completely charmed by this profound, widely educated scholar and brilliant companion, L.A. abruptly interrupted the conversation and turned to Manfred with the confession that his earlier judgment of him had not been correct and he now deeply regretted it.

I remember the hospitable dacha of the Artsimoviches at Zhukovka, always full of young people. We often played volleyball, and L.A. sometimes liked to act like a kid. I remember the smell of smoke from the campfire, where young chefs, students and post-graduate medical students from Uzbekistan, were preparing unbelievably delicious shashlik and plov. L.A. would draw me aside to gossip as he loved to jokingly refer to a cursory exchange of news and telling a new anecdote. At the table L.A. was surrounded by young people; he was almost young himself, with his lively interest in everything around him, in everything popping up in the conversation of his guests.

Nevertheless, one of his most characteristic features in the later years of his life, which stands out more clearly now in retrospect and was not so visible in our relations with the living L.A. was, it seems to me, a kind of suppressed sorrow from recognition of the irreversible nearness of the end. He became much gentler. The sarcasm in his sparkling humor was more and more replaced by lyricism.

He came to love and value everyday domestic life more and more, along with his hobbies like collecting little toy cars and all sorts of bizarre souvenirs from various countries. The Grillots took every opportunity to send L.A. his favorite chocolate-covered cherries "Mon Cheri" from Paris. When he came to our house, L.A. would listen with obvious sadness to music from the collection "Songs of Our Times," 29 records from the years 1916 to 1944, each with twenty-four of the most popular melodies of the year.

Sometimes on various holidays we met the Artsimoviches at the Central

Writers' House. I remember meeting Edmond and Marguerite Grillot at the celebration for the New Year 1971 and also at a party on November 6, 1971.

Right before this party began, we learned that Artsimovich and a group of his colleagues had been awarded the State Prize of the USSR for their work on thermonuclear reactions. The hosts at the Central Writers' House acquired the rare and interesting possibility of hearing the latest news about the most important area in modern science from the mouth of one of its most active creators. As a result, for L.A. most of the evening became a sort of press conference devoted to controlled thermonuclear fusion. His questioners and audience changed continuously: writers generously and regularly gave up their places to their waiting colleagues.

My wife and I rejoiced in the fact that we had infected Lev Andreevich and Nelli Georgievna with our liking for the Pushkin Museum, next to the Scientists' House. We went there together many times for evening readings, listening to the incomparable poetry and prose of Pushkin and compositions about the life of the poet and his friends at the Lyceum. We tried not to miss the presentations of Aleksander Kutepov or the lectures of Natan Eĭdel'man. I remember how once, in 1971 or 1972, Lev Andreevich and I were talking about premonitions. Upon hearing Pushkin's lines "I used to see off every day and every year by thoughts, trying to guess which would become the date of my coming death," I had suddenly decided to look through all the diaries and letters of Pushkin with dates close to the day he died, to try to find traces of premonitions. L.A. became very interested in the search, but we didn't succeed in finding anything.

The last time I got to see L.A. was January 27, 1973. We were going by train to Tbilisi—I to the annual conference on nuclear spectroscopy, and Nelli Georgievna and a friend to give a lecture at the Physics Institute at the invitation of Elevter L. Andronikashvili, and then to visit the Kapitsas in Kislovodsk. We drove from the Artsimovich house to the Kursk station in two cars. Up to the time the train departed, Artsimovich and Artem I. Alikhan'yan stood near the door of the car, both in caps with ear flaps and sheepskin coats, somehow very much alike, as often happens after many years of close friendship.

Lev Andreevich did not look well. He had not recovered as he should have after a heart attack in Poland in April 1972 and was obviously very tired from a recent trip to France.

After returning from Tbilisi I went directly to Poland without stopping in Moscow. I got home only on February 25, Lev Andreevich's birthday. I phoned to congratulate him, but learned that he had already been in the hospital for a few days. He passed away on March 1.

Aleksander Naumovich Frumkin

I think that a scientist, both as a researcher and as a citizen of his country, must work for his descendants, that he must have the future in mind.

A. N. Frumkin

My first impressions of Aleksander Naumovich Frumkin are from the spring of 1942. I recall the noisy corridor of the V. I. Ul'yanov–Lenin Kazan' University, divided by partitions into a number of small cubicles housing the Colloidal-Electrochemical Institute of the USSR Academy of Sciences (CEI), and the door into the inner sanctum: the office of director Aleksander Naumovich in one of these cubicles.

After arriving in Kazan' from blockaded Leningrad, I was hired in April as a laboratory assistant in the laboratory of S. Z. Roginskiĭ, which had already been transferred to the CEI in Kazan' from the Institute of Chemical Physics (ICP), but remained together with other laboratories of the ICP in another building, a five minute walk from the university, close to the Kremlin. It was constantly necessary to run with various missives from the ICP to the CEI, and soon I had my first meeting, rather, collision with Aleksander Naumovich in one of the CEI cubicles. I literally ran into him and was extremely embarrassed, and he somewhat crossly muttered something through clenched teeth and exited into the corridor.

Aleksander Naumovich had passed his forty-seventh year—now this seems quite young to me—but his name was surrounded by an aureole of fame, he was already a scientist famous throughout the world. His manner of looking somewhere off to the side and downward, past the person he was talking to, made one tremble; he seemed to me haughty and unsociable, and I did not at all believe the stories about how simple, clever, and fascinating Aleksander Naumovich was.

Let me jump ahead almost 30 years. Duties connected with my assignment as opponent in a dissertation defense took me to Kazan' for several days, and I spent some hours roaming around the places which had been dear to me since 1942–1943. Of course, I also found myself at the university. And suddenly, remembering the recent 75th birthday of Aleksander Naumovich, I sent him a nostalgia-filled congratulatory telegram. Several days later when we met in Moscow, he was very moved, asked me in detail about Kazan', and reminisced about the war years. The whole time his face wore a kind, sad, and wise smile, and I was again struck by the idea of how wrong my first impressions of him had been.

In Kazan' the Frumkin family lived in the Arbuzov house[1] at No. 8 Shkol'nyĭ Pereulok. The kind hospitality of the hosts harmonized perfectly with the delicacy of the guests. Fairly recently I heard from Boris Aleksandrovich Arbuzov about how embarrassed Aleksander Naumovich was when he had to chase the cat from its usual place on the bed, how hard he had tried to avoid doing this.

The first dissertation defenses which I ever witnessed are also associated with Aleksander Naumovich. One evening at the Institute of Chemical Physics A. B. Nalbandyan was defending his doctoral dissertation, and N. M. Émanuél' was defending his Candidate's dissertation. Aleksander Naumovich was opposed to N. M. Émanuél'. I no longer remember how he appeared that day, or if he even appeared at all or instead his opinion was read out loud, but, combining various such meetings in my memory, I can see the most characteristic pose of Aleksander Naumovich at the table, with head lowered and palm of his hand covering his eyes. I sometimes suspected that he was deliberately asking the defenders tricky questions and barely hiding his laughter while listening to answers which seemed to him absurd. But I soon found out that Aleksander Naumovich was simply shielding his weak eyes from too bright light.

In June, 1943 I was sent to Moscow. The common building housing the CEI and the IGIC (Institute of General and Inorganic Chemistry of the USSR Academy of Sciences) at No. 31 Bol'shaya Kaluzhskaya, left vacant for nearly two years, was rapidly filled. In particular, one of the groups of the hastily created I. V. Kurchatov Laboratory was located here. Aleksander Naumovich was spending more and more time in Moscow, and this by no means made me happy, because his office on the second floor with a large—entending over nearly the entire wall—window onto the courtyard and door covered with shabby black oilcloth was my first Moscow dwelling. I was forced to move in the early mornings from the leather-covered divan in Aleksander Naumovich's office to a room not far off, where I was working on the cata-

---

[1]The house of the well-known Kazan' chemist A. E. Arbuzov. *Ed. note.*

lytic hydrogenation of oleic acid, which formed the topic of my diploma study. N. A. Bakh was opposite me, and Aleksander Naumovich was a frequent guest in her laboratory. Of course, it would happen that exactly at the time of one of his visits a crash was heard coming from my room: the electrolyzer had broken into smithereens. The alarmed Natal'ya Alekseevna ran in asking sympathetically whether or not the explosion was strong. Without any desire at all to make a joke, I confusedly said the first thing that came to mind: "No, not a strong one, just bakh!—and that's it!" Aleksander Naumovich was quite amused.

I remember the very stormy doctoral dissertation defense of O. M. Todes in April, 1944. An organized group of ill-wishing guests strongly opposed the awarding of the Doctoral degree for his dissertation, mainly because many statements of Todes's theory of thermal explosions had been reduced not to analytic expressions, but only to tables of numbers. Aleksander Naumovich incisively ridiculed the hollowness of such criticism, comparing it to the demand that the number π be expressed as a rational fraction. After the happy outcome of the voting, I. A. Kazarnovskiĭ gracefully punned: "Yes, that was a real Todeskampf" (German for 'death struggle').

In the fall of 1944 I became a graduate student at the ICP, and encounters with Aleksander Naumovich became much rarer, especially after 1947, when after my dissertation defense I was forced to abruptly change my topic of research (for more than 10 years afterward I worked solely on nuclear physics). Before long, it turned out that Aleksander Naumovich was one of the main figures accused of cosmopolitanism and grovelling before the West (other targets were N. N. Semenov and P. L. Kapitsa; the latter was particularly vulnerable to any sort of criticism, as since August, 1946 he had fallen into disgrace, had been banned from his institute, and forced to live in a dacha).

There was no shortage of targets both in the chemistry department and the physics department of Moscow State University at that time. Among the common types of damning evidence were frequent references to the work of a Soviet scientist in the foreign scientific literature, translation of the scientist's works into foreign languages, and, in general, admiration or even simply respectful treatment of the scientist and appreciation of his scientific work in the West, and A. N. Frumkin was not lacking in any of this.

At the end of the 1940s (I no longer remember the exact date) Aleksander Naumovich was removed from his post as Director of the Institute of Physical Chemistry. The repercussions of these events reached me more and more strongly in the summer months at our family's dacha in Lutsino near Zvenigorod, where beginning in 1949–1950 I had the happiness of interacting with such persons as Aleksander Naumovich and his close friend and admirer Pyotr Aleksandrovich Rehbinder.

Among the more pleasant traditions of the dacha seasons at Lutsino were the July celebrations of the name-day of Pyotr Aleksandrovich. The noisy, life-loving celebrant excellently combined the role of hero of the festivities with the duties of master of ceremonies. The merry toasts never stopped, and humorous verses specially requested by Pyotr Aleksandrovich and Aleksander Naumovich on current topics of interest to the guests resounded. They often requested verses by my closest friend, the excellent theoretical physicist and no less remarkable poet A. S. Kompaneets, and I knew almost all of them by heart. How pleasant it was when these verses met with an approving reaction from Pyotr Aleksandrovich and Aleksander Naumovich; one extroverted and noisy, the other restrained, barely showing anything (a slight smile, glints of laughter in his half-closed eyes), but no less happy than the other. Aleksander Naumovich loved to play tricks on the boisterous, vivacious Pyotr Aleksandrovich. He had very warm relations with Elena Evgen'evna Rehbinder, and it was usually he who raised a toast to the health of the mistress of the house.

The long walks with Aleksander Naumovich and Pyotr Aleksandrovich in the surrounding woods were unforgettable. Both were fond of "doggies," so every walk was accompanied by continuous cries and sometimes by searches for incessantly disappearing dogs. Because of his peaceful nature and small size, Frumkin's Foma could not really frighten anyone, but the succession of dogs belonging to Pyotr Aleksandrovich were very large, so that passersby were often terrified by them. At each such encounter Pyotr Aleksandrovich would begin to cautiously call the dog to himself while at the same time extolling its remarkably peaceful nature. Such encounters often reminded us of the "dog order" at our dacha settlement, and we would delight in newly minted formulations of it, for example, beginning with the words: "In view of the fact that the dog of Academician B. named Jack, barking at passersby, threatens to jump over the fence, which is practically within his capabilities... ."

Our conversations on these walks touched upon everything: scientific news, political events, gossip about common acquaintances, and what used to be referred to as "society news." I always marvelled at how well informed Aleksander Naumovich was not only about topics directly related to his interests, but also about a wide range of topics currently in fashion in elementary particle and nuclear physics. With some bias he would ask about what some or other scientist meant by his "Hamburg score," or how much his official recognition was consistent with his real worth.

A close working relationship between Aleksander Naumovich and myself was established after fate led us to the Bureau of the Division of General and Technical Chemistry of the USSR Academy of Sciences created in 1963. A. N. Nesmeyanov was chosen as Academic Secretary of the Division.

Aleksander Naumovich became one of his assistants and directed the coordination of all the activity of the Division in physical chemistry. Among the problems Aleksander Naumovich dealt with were the creation of scientific committees to deal with various problems in physical chemistry and the supervision of their work, organization of the scientific sessions of the Division, and keeping track of the plans and accounts regarding topics in physical chemistry.

Aleksander Naumovich dealt with his new duties with his customary extreme conscientiousness. Each interaction with him was deeply satisfying owing to the combination of human warmth and his exceptionally rich and broad scientific experience. His advice during the creation in 1963 of the scientific council on high-energy chemistry was invaluable. As is well known, Aleksander Naumovich was one of the pioneers of Soviet radiation-chemistry research, which began in the mid-1940s. Until the end of his life, Aleksander Naumovich retained his interest in this field, which he viewed as related to electrochemistry (after all, radiation chemistry is also to a large degree electron chemistry). Owing to his efforts, the Institute of Electrochemistry became one of our most important centers for radiation chemistry.

In the fall of 1963 Aleksander Naumovich and I prepared the first scientific meeting of the new Division, which was entirely devoted to physical chemistry. I remember how happy he was with the success of this meeting, the breadth of the topics dealt with in it, and the successful combination of presentations by both venerable and novice speakers.

The clarity of Aleksander Naumovich's thinking and his ability to compare and analyze seemingly unrelated facts from all sides and to cleverly combine the possibilities of different methods for solving problems of interest to him always impressed me. For example, at this meeting, after listening to the report on our research into the mechanism of oxidation-reduction reactions in solutions by a new method based on observations of positron annihilation, Aleksander Naumovich immediately saw and evaluated the possibility of an independent way of testing his earlier idea that electron transfer between ions occurs in a solution not via long-range tunneling, but inside an intermediate complex.

Many years later our review of positronium oxidation in solutions appeared in the memorial issue of the journal *Électrokhimiya* dedicated to the memory of Aleksander Naumovich.[2]

Among the remarkable features of Aleksander Naumovich which I would like to note particularly in these brief reminiscences were his firmness and courage, his refusal to compromise, and his perseverance.

---

[2]V. M. Byakov, V. I. Gol'danskiĭ, and V. N. Shantarovich, "Energy of positron hydration and positronium oxidation reactions," *Élektrokhimiya* **13**, No. 6, 804 (1977). *Ed. note.*

Toward the end of his life Aleksander Naumovich had to bear many burdens. He lost Amaliya Davydovna, with whom he had spent many decades of conjugal life and shared most of his interests, likes, and dislikes. The glaucoma which had long tormented him became worse, and he was forced to spend more and more time in the hospital. Right after his 75th birthday difficulties arose with his reappointment as institute director for a new term. His deep disappointment with one of his closest students weighed on him a great deal. But no difficulty could overpower him. As always he worked hard on his beloved science, he devoted all his energy to the institute which he had founded and directed, and he fiercely defended his most highly regarded candidates in elections to the Academy of Sciences. And his lifestyle remained nearly unchanged: an evening nap before working on articles and books until long after midnight, trips to the dacha from Friday until Sunday evening, walks on the paths around Lutsino. P. A. Rehbinder had passed away, and now there were usually only two of us on the walks. Sometimes we were joined by Émiliya Georgievna, who had become a strong support for Aleksander Naumovich and devotedly shared with him all the joys and sorrows of this last period of his life.

The scientific prowess of Aleksander Naumovich was as impressive as ever. His brilliant report was the most noteworthy event at the Eleventh Mendeleev Congress on General and Applied Chemistry in Alma-Ata in September, 1975. The audience of thousands gave a standing ovation to the most venerable of Soviet electrochemists after his concluding words about Aphrodite, born of sea foam, which after all is nothing but a cross-linked electrolyte.

At various jubilees Aleksander Naumovich loved to offer a joking toast about age. He had discovered in some book, apparently an encyclopedia, a classification of ages according to which people from the age of 50 to 80 are in their maturity and pass through two stages of being elderly, then for five years (80–85) they pass through ill-fated old age, after which they can be classified as long-lived. "Therefore," added Aleksander Naumovich at the jubilees of P. L. Kapitsa and N. N. Semenov, "let us drink to our celebrant's successful passage through a few years of being an old man, and then all will go well."

In the course of quickly—no, too quickly—passing years and decades there were two jubilees for Aleksander Naumovich himself.

In October 1965 the packed auditorium, an ampitheater, of the chemistry department of Moscow State University was the scene of the celebration of his seventieth birthday. Short speeches of welcome alternated with merry jokes, each revealing the love and deep esteem for the celebrant. There were funny gifts and film clips (including the touristic exploits of Aleksander

Naumovich—from when? I don't remember exactly). Then a supper with colleagues, where N. N. Semenov, who arrived last, excitedly reported to a storm of applause the news—he was late because he had been waiting for it—that the celebrant had been awarded the honor of Hero of Socialist Labor.

And then there was October 1975, when Aleksander Naumovich's faithful assistant and secretary, Aleksandra Sergeevna, greeted the disorderly flow of guests in the anteroom of his office in the Electrochemistry Institute, and then presented the guests, already broken into small groups, properly to the celebrant. However, in his office the propriety suddenly vanished, it was so joyful to see the lively, smiling Aleksander Naumovich, wineglass in hand, to listen to his jokes, to embrace and kiss the hero of the celebration. Taking me aside, Aleksander Naumovich laughingly told me the story of how that very morning he had found himself in an embarrassing situation. He had been urgently called to see the Vice President of the USSR Academy of Sciences. He immediately set off for the Presidium of the Academy of Sciences in the complete certainty that he had been invited to be congratulated on his jubilee, although he was certainly somewhat surprised that he had been asked to go to them rather than them coming to him. Entering the office of the Vice President with his words of thanks for the congratulations already prepared, Aleksander Naumovich learned that there was a completely different reason for his invitation: he was asked to sign a routine counter-Sakharov joint letter from a group of Academicians to a newspaper. It had to be seen and heard how Aleksander Naumovich fully appreciated the humor of the situation, with what pleasure and detail he described his "jubilee" visit to the Presidium. And to that day Aleksander Naumovich continually repeated his favorite joke about age.

Alas, less than a year had passed before thousands of comrades, friends, and students again came to the University to see Aleksander Naumovich, this time to say goodbye to him.

Looking at his portrait beside his grave, at the barely smiling, kind, wise face of Aleksander Naumovich which has become so dear over long years of acquaintance and friendship, I sadly think of how I am forever separated from one of the most interesting and remarkable people with whom fate has allowed me to become acquainted in the course of my life.

On the 70th Birthday of Vitaly Lazarevich Ginzburg

Not wanting any stories about his personal life or his work to appear in print, Mayakovsky emphasized the fact that he was a poet, and was interesting precisely for that reason (and no other). However, in fulfilling the request to write about Vitaly Lazarevich Ginzburg on the occasion of his 70th birthday, I intend to write little about his scientific accomplishments; these can better be described by the scientists working with him (among whom are many of our most eminent physicists) and his comrades-in-arms. I would instead like to sketch some of his wonderful human qualities. This is difficult, particularly when writing about someone who himself is the author of a number of lively and interesting reminiscences.

A recent review in the magazine *Nature* (October 10, 1985) of the English translation of the book *On Physics and Astrophysics* by Ginzburg begins with the sad words: "The era of specialization has now advanced very far. The last scientists whom can be referred to as 'physicists' in general, rather than 'solid-state physicists' or 'high-energy physicists' are getting close to retirement." The author of the English review correctly sees Ginzburg as one of the last physics encyclopedists who has survived to our times. And this feature is characteristic to the highest degree of Vitaly Lazarevich (VL) not only in science, but also in life. In general, the book *On Physics and Astrophysics* serves as an excellent portrait of its author not only as a scientist, but also as a human being, with its clear partiality to subjects in keeping with his tastes and interests and the clear unwillingness to discuss "distasteful" problems, with its strong and open (although at times also unaccountable) sympathies and antipathies. VL characteristically has direct and open relationships with other people, relationships which can be termed childlike, in the good sense of this word. He is warm, even passionate, in everything—in lectures, storytelling, sports, or watching soccer or hockey games on television, when he literally jumps up in the chair and yells and stamps his feet in annoyance. His intonations with stress on certain words and his gestures are extremely infectious. After spending some time in his company, people unconsciously begin imitating his speech and gestures.

VL is brilliant at the blackboard; I will not undertake to describe his special charm as lecturer or speaker, he must be seen and heard in person.

VL's interventions at assemblies of the Academy of Sciences usually attract general attention and give rise to arguments. Their special feature is always a perceptible passion: the orator's deep conviction of the validity of his words and the extreme necessity of what he proposes.

This conviction strongly increases VL's effectiveness as a force behind elections to the Academy. The picture of a sports fan watching television is only a weak analogy for the enormous amount of emotion with which VL greets each such election. Rare talent is happily combined with a no less rare love of work in VL. However, one can hardly speak of love of work when work for a human being is a true pleasure, an inexhaustible source of positive feelings. After an intense week of work, from Friday evening until Monday morning, in any sort of weather, he—not only willingly but with great enthusiasm—shuts himself up in his dacha, sometimes in isolation, and reads, thinks, and writes. And then there are also his excursions, on foot and on skis, and he recently with his own hands sawed up the huge trunk of a fallen tree. When he goes off somewhere for vacation, VL takes along a stack of magazines—*Scientific American*, *Nature*, and *Priroda*—and avidly reads them from cover to cover. With lively interest he questions his acquaintances about news of their own work and in related areas. "What does the scientist work on?" he asks, smiling with his characteristic ironic intonation, with a gleam in his eye, ready to listen for hours to someone else and to tell about his own work, and sometimes to improvise a search for an explanation of some new fact. The enthusiasm with which he listens to the stories of experimentalists has sometimes even let down either him or, in the final analysis, the same experimentalists; it has happened that VL quickly found and published the theory of some new phenomenon supposedly already discovered experimentally, but which in fact has yet to be discovered.

"Not a day without a line," is the writer's slogan completely adopted by VL, and this in combination with his passion and lively pleasure in all things close to his heart sometimes makes it necessary to beg him not to spill his emotions onto paper. Such was the case, for example, when he was indignant about the miserable existence of a dwarf hippopotamus in a certain zoo which we had visited together, and he strongly wanted to write to some important newspaper complaining about it.

VL has an excellent sense of humor. All his friends remember, for example, his favorite skit: a radio broadcast for children on the content of some newspaper editorial, particularly from "Pravda". VL reads the article in a thin, whining voice, inserting the word "children" several times in each line of text.

VL is an ardent propagandist of science, an organizer for many years of all the scientific sessions of the Division of General Physics and Astronomy of the USSR Academy of Sciences, and the supervisor of one of our oldest and most famous physics seminars. His fruitful activities in the philosophy of the natural sciences is widely known.

The scientific achievements of VL are widely acclaimed throughout the world. He is a foreign member of many illustrious academies of science, doctor *honoris causa* of a number of universities, and winner of the Lenin Prize and the State Prize of the USSR.

For all this VL is a complete stranger to any servility, whether directed toward him or expected from him. He is completely even-handed in his dealings with people: in praise or blame, admiration or indignation, "high" and "low." Therefore, the accurate characterization of an "isotropic boor" given by him to a common acquaintance sounded rather like a sort of wondering praise: even boorishness can be pardoned if it is manifested in the same degree to both subordinates and superiors.

It is almost never possible to refer to VL using the two words "Academician Ginzburg." He therefore had some difficulty getting cars from the motor depot at the Academy, until the people there got used to the existence of such an academician and began calling him by only his family name.

Incidentally, in connection with this I am reminded of an excellent toast offered to another of our scientists upon his election to the Academy. "Now you will sign papers 'Academician So-and-So'," one of our friends said, rising from the table. "I raise my glass to propose that the second of these words will always mean much more than the first."

Physicists everywhere understand the words "Academician Ginzburg" in just this way.

Twenty years ago, speaking to a huge crowd at the Lebedev Institute, Boris M. Bolotovskiĭ ended his very witty verses for the 50-year jubilee of VL with these lines:

> Our country repays you with kindness and concern,
> Now you do not have a life, but a paradise.
> Tell us—why on earth
> Do you still want to write and lecture?
> You are not a child, your papers are innumerable,
> You are an Academician, everyone recognizes you,
> It's time to stop your scientific work,
> Or else people will simply begin to laugh at you.

There is no doubt that even in his hundredth year (grant it, God!) Vitaly Lazarevich will not slacken a bit in his drive to "write and lecture," because life and work are synonymous for him.

In the Kaleidoscope of Memory
(Lev Davidovich Landau)

Dau usually sees me in his small, cozy room on the upper floor of the apartment in the courtyard of the Institute of Physical Problems, in his favorite pose: semi-reclining on a divan, leaning on his left elbow, resting his back on the rug covering the wall. His right hand actively participates in the conversation; it is bent at the elbow, the palm of his hand thrust forward and stressing his ideas with energetic motions. Sometimes his speech contains sarcastic tones, at other times he speaks almost in a singsong, especially when he is celebrating his victory in some argument or speaking on one of his favorite topics, when the text is almost canonical (we refer to these texts as "Dau's records"). How did it happen that we did not record even a single monologue of Dau on tape or at least on paper, but still the details are fresh in our memories!

Our interaction began in the winter of 1942–43 in Kazan', in a youthful crowd, in the home of future Academician Georgii Konstantinovich (then simply Zorik) Skryabin and his wife Irina Borisovna, Ira. I do not remember who invited me there and introduced me to these hosts or the details of this evening. I only remember that Lev Davidovich was at the center of attention, that he talked a lot in a lively manner and behaved very simply, just like anyone else. I also remember the trip home together, walking through Kazan' in the dark via the "Heights," through little parks and valleys, from Bolshoi Krasnoi to Kalinin street. I was trying to find out how much mathematics and theoretical physics a chemist needed to know, whether the university courses were sufficient, and what more was necessary. Lev Davidovich answered that the theory of functions of a complex variable was not so important, but vector analysis was absolutely essential, and he gave many specific examples of this from physical chemistry. He spoke in just as great detail about statistical physics, perhaps because E. M. Lifshits was then giving this course to all aspiring chemists in one of the auditoriums of Kazan' University, in whose buildings in 1941–43 were located our most famous physics and chemistry institutes, in extremely crowded but friendly conditions.

Our acquaintance continued in Moscow, where I moved in the summer of 1943. The building in the courtyard in the Institute of Physical Problems was rapidly filled with coworkers returning from Kazan'. I used to drop by my relative A. I. Shal'nikov, with whom Khariton lived for a short time, and more and more often met Lev Davidovich, who soon asked me to simply call him Dau.

By the end of the war the people at the Institute of Physical Problems had begun to gather frequently in the institute hall to dance to the music of a radio-gramophone. These dances were presided over by Oleg Pisarzhevskiĭ, the abstractor of P. L. Kapitsa and future author of *Mendeleev* and *Fersman* (books in the series *Lives of Remarkable People*). Although Dau did not dance (at most he sometimes half-jokingly stamped in time to the music), he almost always also appeared at these gatherings in the institute and sat in a corner somewhere, closely surrounded by male and (quite important!) female acquaintances. His stories and lectures—already reminiscent of the "Dau records"—successfully competed with the dance records, many of which the company Melodiya has recently reissued for us. He really loved those dance evenings, and often pestered me to go with him to them, either at the Institute of Chemical Physics in the late 1940s, at the Lebedev in the 1950s, or in later years, up to the catastrophe, at the home of my new friends from Koktebel'. His appearance was always an event, and he himself openly enjoyed all the sights of the evening, the festive music, the many beautiful and lively faces all around. In addition, everyone around him enjoyed an unexpected meeting with Dau and the chance to speak and interact with him.

We began to see each other especially often after 1946, when I settled in the courtyard of the Institute of Chemical Physics as a close neighbor to Dau. In the winter of that year a couple of times I heard surprisingly clear popular lectures given by Dau on the most arresting topic of that time: the atomic bomb. In August P. L. Kapitsa fell from favor, and this provided a new topic for Dau.

I remember an evening party at the Institute of Chemical Physics at the end of 1946 celebrating the election of Yu. B. Khariton and Ya. B. Zel'dovich as Corresponding Members of the USSR Academy of Sciences. The humorous verses I read that evening about those present included the lines:

> "What is Dau thinking about, always smiling in contemplation,
> probably he has seen a lassie, very deserving of attention."

At the same elections Dau himself became, as is well known, an Academician, bypassing the stage of Corresponding Member. Arriving to congratulate him, I recalled the well-known English expression "the gentle art of making enemies." Dau delivered witty, brief, accurate characterizations of several of our physicists, sometimes annihilating them in pairs, other times in whole series. "Dau," I said to him, "there is no doubt that you are our best physicist. Why then is it not a greater honor for you to be the best among the best than to be the best in this mediocre series?" Dau laughed and "changed the record."

Some fragmentary reminiscences from the next few years.

Dau's love for classification. I will not in detail repeat here the widely known examples quoted in books about him and probably already described in the present volume of reminiscences. For example, theoreticians of the top rank like Newton and Einstein, and theoreticians of the next lower rank—twelve plus (posthumously) a thirteenth, Fermi. Classification of men according to what they value most in women (kindheartedness, figure, beauty, etc.; with condemnation he classified me with those who value kindheartedness). And then, according to Landau, the five classes of "public places," in order of decreasing quality: office, business, shop, tavern, brothel. Dau regarded his own institute very highly, almost in the highest class, and he treated our Institute of Chemical Physics as a shop, mainly for employing the excess number of scientific workers, who moreover were not always, in his opinion, of very high quality.

One of his favorite ways of characterizing a person under whose wing various scoundrels were protected was as a "trash-lover." This pertained also to those who voted for unworthy candidates in elections to the Academy.

Several times Dau joined me and my wife, with whom he had a warm friendship, on ski treks to Uzkoe. Even though I am a very mediocre skier (not without reason did Ya. B. Zel'dovich, watching my wife and I ski down a mountain four months before the birth of our youngest son, say that he got the impression that of the two of us it was I who was pregnant), Dau did not even reach my level. He didn't ski for amusement or exercise; he skied only as a means of transportation, and nevertheless derived great satisfaction from it.

Some more scraps of memories.

Dau came to me on my birthday, when I reached the wonderful age of 27. "Well," he declared, "this is the last time Vitya Gol'danskiĭ (he often called me this, with the stress on the last syllable) will celebrate a birthday in the series n^n, where n is an integer." Well, of course, no one has yet managed to live to the age of 4^4=256.

Dau loved Simonov, and knew some of his verses by heart.

In the summer of 1951 before our trip, which required passes, to Novyĭ

Afon he phoned to say goodbye, and asked us to procure, no matter what, the small pamphlet by Dmitriĭ Gulia "Concerning My Book *The History of Abkhazia*," as it could be very useful in case one in future had to repent for something.

Indeed, this is an excellent example of searching for one's own identity. It appears that the famous popular poet of Abhazia was under severe pressure, so that from his pen flowed lines that I always remember, the lines with which the brochure begins: "For a host of reasons, from my pen in years long past came a pseudoscientific, antimarxist book, in which my premises and conclusions are confused." And several lines later: "Let us follow the author further. Here is another example of my groundless considerations."

Dau would read these phrases with exaggerated charm and in a singsong voice, waving his hand in time with them.

Beginning, I think, in 1952 we, for several years in a row, celebrated the New Year with a large group, successively at various homes in the buildings belonging to the Institute of Chemical Physics and the Institute of Physical Problems, and in the evening of January 1 we would get together to celebrate the "stale name-days." The other traditional gatherings were held on birthdays. We played charades, in which the superstar (as in poetic works on topics concerning physics and physicists) was A. S. Kompaneets, the eldest of Dau's students, the first to have passed his theoretical minimum. Dau burst into loud laughter when for the syllable "rov"[1] Kompaneets, in tight-fitting long underwear, climbed up a stepladder and attacked a fortress—a cupboard, and for the syllable "AN" depicted Academician T. D. Lysenko reporting on superabundant harvests and superproduction of milk and describing his "scientific" methods.

For New Year's 1956 Kompaneets and I wrote a play entitled "Day of a scientific man" and performed it to an audience on the night of January 1 in Dau's apartment.

Meanwhile, the times were changing significantly (in our play this found expression in the scene entitled "Rumors"). The revival of international physics meetings was approaching, along with the appearance of Dau at Moscow State University, where for a long time the trends in physics had been completely unacceptable to him.

In the spring of 1956 the Lebedev Institute was the setting for a large international conference on high-energy physics, attended by stars of the first rank, among them eight (four Soviet and four American) future Nobel prize winners.

Although I had clearly understood how talented Dau was, I was still hap-

[1]Meaning "ditch" or "moat" in Russian.—*Transl. note.*

pily impressed by the extraordinary reverence that the foreign guests had for him. All he had to do was rise from his seat or raise his voice in a question or comment for the densely packed hall to become quiet and each guest give him his full attention. The still quite young-looking Murray Gell-Mann literally did not take his admiring eyes off Dau and was completely enraptured with him. The private reception held for the foreigners by Dau was very warm and friendly. That spring, the prologue to which was the Twentieth Congress of the CPSU, was characterized by a great merriment, joking, and high hopes.

I remember also other international conferences in which Dau participated: Kiev in the summer of 1959, and Moscow in the summer of 1960.

The setting was the hallway of the hotel in which we were all staying during the Kiev conference. Heisenberg and Dau were speaking quietly about something, with a group of respectful and curious listeners not far off. Alvarez approached the great ones and alternately took each off to the side and conducted a test. He revealed one after another the numbers in a column: 1000, 40, 1000, 30, 1000, 20, 1000, 10, and then asked each to quickly give the sum. A thousand, a thousand forty, two thousand forty, two thousand seventy, three thousand seventy, three thousand ninety, four thousand ninety .... But instead of a total of four thousand one hundred, everyone for some reason instantly answers five thousand. That time the trick was especially impressive: both Heisenberg and Dau made the error, and Alvarez, satisfied, went off with his book of problems to other groups of people.

A brief speech by Dau from his seat, in which he, barely able to keep himself from laughing, literally destroyed one of the leading speakers at the Moscow conference on nuclear reactions is to this day one of the highlights of Mössbauer's favorite story about his first visit to Moscow.

Here I shall not recall the celebration of Dau's 50th birthday in the Institute of Physical Problems; I think that this event has been described well enough, since the eyewitnesses and active participants in this jubilee numbered in the hundreds. Thousands of viewers remember Archimedes Day on the steps of the physics department of Moscow State University in May, 1961 followed by a party in the theater, where the guests of honor were Niels Bohr and Dau.

Let me therefore turn to more "intimate" examples, now directly pertaining to physics.

In the winter of 1954 owing to the initiative of Ya. A. Smorodinskiĭ our nuclear experimentalists received a remarkable gift: a course of lectures on nuclear theory given by L. D. Landau at the I. V. Kurchatov Institute of Atomic Energy, and published in 1955 as a book by L. D. Landau and Ya. A. Smorodinskiĭ. This course played a very important role in the development of Soviet nuclear science as a fundamental science, and drew the attention of a wide range of physicists working in other fields.

In the fall of 1954 M. I. Podgoretskiĭ and I asked Dau to submit to the journal *Doklady Akademii Nauk SSSR* our article on the use of correlation functions to study the mechanism of nuclear transformations. In this article we showed that the well known coincidence method is actually only a special case of the much more general correlation method, which is moreover free from any constraints on the resolving time.

Dau immediately went to the heart of the problem, which we had spent many weeks working on, and demanded that we add to the article the derivation of the formulas for the dispersion of the correlation functions, i.e., that we show what are the statistical errors of the generalized method. After managing to complete this assignment, we clearly saw that only then did the article really become definitive. And to this day I take pride in the fact that the title of this old article is followed by the line: "Submitted by Academician L. D. Landau, 9 October, 1954."

At the end of 1956 or beginning of 1957 I went to Dau and found him unusually animated. He took me upstairs and enthusiastically told me in detail about his idea of conservation of combined parity. I probably never again heard him speak with such youthful enthusiasm about his "cooking," about how he came up with the idea that the invariance of physical laws under the transformation of left into right (he showed the three coordinate axes by the fingers of both hands pointing towards each other) requires that left-handed electrons become right-handed positrons. At times he became thoughtful, withdrew into himself, and fell silent, and then again he would continue his enthusiastic story. He did not write down a single formula, sometimes he only stated one sketchily, cursorily, like a long familiar text.

Here is an example of the Landau "mathematics in everyday life." In the car on the way to Uzkoe, my wife and I were telling Dau about an actor who had performed the following trick the previous evening in the Writers' Home. Before him was a large vertical abacus-like object with three horizontal wires, and on the left sides of these wires were respectively 3, 5, and 7 rollers. Two people were to take turns moving any number of rollers from the left to the right on any one wire at a time. Whoever moves the last roller, leaving none for the other player, wins. Dau meditated, and the conversation stopped. When we were unloading our skis in Uzkoe, he gave a general solution of the problem. If the game is played correctly, the player who starts always wins. In his first turn he must move one roller of the set of seven, and after that he must play such that after each of his turns in the sum over the three wires there is an even number of 1s in each of the three binary digits. In response to puzzled questions from my wife, Dau explained what a binary number is, what it is needed for, and why 3, 5, and 7 are 11, 101, and 111. Certainly, none of the artists at the Writers' Home could have imagined that the rules of this game had such a theoretical foundation!

The years passed, and now in retrospect much has merged together, and it is often difficult to say what happened earlier and what later. The fatal Sunday of January 7, 1962 approached.

Dau phoned us four days before. My wife answered the phone. She thought that he wanted to speak to me, but Dau was actually phoning her, not on business, but to wish her a Happy New Year. The subsequent events have forever imprinted in my memory the greeting that Lev Davidovich and his playfully festive tone, although there was nothing at all unusual in these greetings: "Now, when the peoples of Africa are successfully fighting for their freedom, I wish that you will acquire independence and freedom in the New Year," and more of the usual "Dau record" about how it was necessary for us to rest separately, how one should not travel to Tula with one's samovar... .

We were to meet "that week" and planned to visit friends together.

On Sunday evening N. N. Semenov received an alarmed phone call from A. I. Shal'nikov urgently searching for pure thiourea, required for emergency neurosurgery. I remember a box with the address: "N 1. Prof. Lev Landau c/o BEA Medical Officer, BEA Medical Centre, London Airport" from the first package of medicine urgently sent for Dau by Maxwell at Pergamon Press by the first flight from London. The epic struggle for Dau's life had begun.

In the fall at the Institute of Neurosurgery I was for the first time allowed to see him in his room, where he lay in bed, dozing, unusually alienated, with a somehow unfamiliar face. A. I. Shal'nikov and another man unfamiliar to me were also in the room. Soon the latter left, and I asked Shal'nikov who he was. Shal'nikov gave his surname, and then we heard the weak voice of Dau saying, "A very weak physicist," in criticism of the visitor. How happy I was—the smoker was alive! It seemed to me that Dau would again become like before, that everything would return to the way it had been. Alas, these hopes were in vain. Dau completely stopped talking about physics, he was nearly indifferent to his surroundings, and he constantly complained of pain in his leg, of his loss of memory. His memory was somehow selective, for example, he could speak English as fluently as before, but many things had simply disappeared. He had completely lost the ability to learn anything new. For example, Ya. A. Smorodinskiĭ and I several times told him the news about the two types of neutrino, electron and muon, and each time he greeted this as though he were hearing it for the first time. But also he sometimes expressed very accurate thoughts and characterizations, and then his listeners saw a spark of the former Dau.

Particularly typical of his new state, the tragic situation of the fettered Prometheus, was our conversation in the hospital of the USSR Academy of Sciences in October, 1962 on the day which would have brought great happiness to the healthy Dau, the day he won the Nobel Prize.

"Dau, I congratulate you on winning the Nobel Prize," I said, coming up to him.

"It should be pronounced Nobél, not Nóbel," answered Dau and asked who else had received the prize. I named Steinbeck.

"I don't like Steinbeck."

"What do you mean? Since Faulkner and Hemingway died he is the greatest American writer."

"Is Hemingway really dead? Of course, that was no longer in my time."

These words of Dau literally rent my soul. After a pause, I continued:

"No, Dau, he died in the summer of last year. I very much loved Hemingway."

"Why 'loved'? You have in mind not him presonally, but his works. You should say 'love.' "

After a pause, Dau continued:

"How is Mila? How is your son?"

"Which one, Dau? We have two."

"Two? I don't remember. I only remember one. I don't remember many things, for example, I have completely forgotten how my parents died."

There were five more years of this "different life" still ahead for Dau, but it is sad and difficult to write about this. We always remembered Dau as he was before his first death, and will always be grateful to fate for allowing our lives to somehow intersect.

Aleksander Nikolaevich Nesmeyanov

January, 1943. Kazan'. Not without fear, I go to Aleksander Nikolaevich to take the exam on the theoretical foundations of organic chemistry. The setting is Aleksander Nikolaevich's laboratory. After getting the questions I begin to write, and in the meantime he goes around to several of his colleagues, asking them about their work and giving advice on how to do it. It is like an orchestra rehearsal, where the conductor works out the different parts of the score with the musicians responsible for each part. This was a very interesting show, and I was involuntarily distracted from my work to watch and listen to Aleksander Nikolaevich. Finally, he brought over a chair and sat down next to me. He asked almost nothing related to the actual questions, but digressed to related topics, checking more on my understanding of the course than my knowledge of it.

That A seemed to me more difficult to earn and more valuable than all the others I received!

... Fall 1945. Uzkoe. An intense game of skittles is going on. The two teams are led by two Academicians—Aleksander Nikolaevich Nesmeyanov and Vladimir Mikhaĭlovich Rodionov. Aleksander Nikolaevich loves skittles. He plays with great passion, constantly making "sporting" exclamations typical of skittle players like, "And now we're going to smash your little engine into dust!" Thirteen years later when we were playing skittles on an asphalt road in Lutsino, these exclamations were of a different nature: the threats of destruction were addressed not to the other skittle players, but to the specifically named, zealous persecutors of resonance theory in the then ongoing (sad memory!) controversy over the existence of a resonance in organic chemistry.

... April, 1957. Moscow, the President's office. Attending a reception with

Aleksander Nikolaevich are six people: everyone in our proposed delegation led by N. N. Bogolyubov to the Rochester Conference on High Energy Physics in the United States. After much work and numerous "to be or not to be" swings of the pendulum our trip had been cancelled. Our last hope was Aleksander Nikolaevich. But he knew better than we did how obdurate were the opponents of travel in those days, and he only offered us the wise advice: "Believe in the sine function!" How many times after that day did I repeat these words of Aleksander Nikolaevich at difficult moments in life, and how often have they proved true! Incidentally, their first confirmation came quite soon, after only a few months, when the question arose of whether or not I would attend the Gordon Conference on nuclear chemistry in the United States in June of the same year, 1957; the answer was yes.

... Spring 1961. Moscow, the Prague restaurant. There is a reception in honor of Niels Bohr, the last one held by Aleksander Nikolaevich as president of the USSR Academy of Sciences. Of course, he already knew of his imminent retirement, but, as usual, he was calm, happy, sometimes ironic, giving an elegant tone to an exceptionally friendly and warm meeting at which all the greatest scientific stars of our country were present. Now, alas, almost none of them is living.

In the room a rumor circulated of a recent conversation between Aleksander Nikolaevich and N. S. Khrushchev. The latter, dissatisfied with the Academy of Sciences, had allegedly announced his intention of dissolving it, and Aleksander Nikolaevich (who had bluntly refused to support T. D. Lysenko) had answered: "Well, so Peter the Great founded the Academy, and you will shut it down!"

I do not know if this rumor is true, but, as Niels Bohr said, "A good story need not be literally true to be the truth."

... For many years, beginning in 1963 when Aleksander Nikolaevich became Academic Secretary of the Division of General and Technical Chemistry of the USSR Academy of Sciences, I was the delighted witness of how he conducted both administrative meetings and general meetings of the Division—in a lively, elegant manner, with humor, and at the same time very business-like. As a rule, he spoke very directly, sometimes sharply, not considering it necessary to hide his opinion when it was negative, say, when the discussion was about whether or not to give a high award to someone in his opinion undeserving of it. After all, for an academic secretary it is so easy to vote "in favor" with all the others, and then to torpedo this yes vote at the request of a higher authority. However, Aleksander Nikolaevich considered such tricks beneath his dignity.

... At the general elections to the Academy of Sciences in 1974, in a Division meeting, Aleksander Nikolaevich, supporting the candidacy of N. A.

Platé as Corresponding Member of the Academy of Sciences, after a laudatory speech about the candidate threw out the joking aside: "And his genes are good—he is the grandson of Zelinskiĭ!"

Scientist, Thinker, Humanist

(Andrei Dmitrievich Sakharov)

Even in 1950, when this very quiet and shy young man first caught the author's attention, a bright future was seen for Andrei Dmitrievich Sakharov.

I cannot say that I knew Andrei Dmitrievich Sakharov well, so this will be a view of him from off to the side, so to speak.

It was 1950, in Dubna, when I first saw Sakharov. I was working at the first of the Dubna accelerators at the time. One day, while a management team headed by Igor V. Kurchatov was visiting the place to get a look at the experimental results, a very quiet and shy young man caught my attention. When I asked some acquaintances who he was, I was told, "Sakharov, candidate of sciences from the Lebedev Physics Institute." I was advised to remember his name, since even at the time a very bright future was seen for him.

At a seminar, the leaders of the various groups working at the accelerator spoke about what they were doing. I was one of these speakers, representing my colleagues in the Institute of Chemical Physics. After the seminar, Sakharov walked over and said some kind words about our work, but I sensed that he was saying much the same thing to almost all the speakers. That was the style of his generosity: Very polite and attentive, Andrei Dmitrievich strived to emphasize the positive aspects of his impressions.

In 1953 I heard that Sakharov had defended his doctoral dissertation. The defense had been specially organized on short notice. The reason for the hurry was that elections to the Academy of Sciences were to be held that same year, and Sakharov was to be promoted to the rank of full member of the academy. Although no one doubted that eventually he would go far above the doctoral level, it was important to observe the formality: It would have been a bit awkward to elect a person without a doctorate to the top rank of academician. To the best of my memory he received 23 of the 23 votes from his own department.

In 1961 our paths crossed at a Crimean resort, where Andrei Dmitrievich was taking a vacation with his children. Beyond the very ordinary conversations that all the guests had, I did not talk with him. I was not alone in this regard: I do not recall that my father-in-law, the physicist Nikolaĭ Nikolaevich Semenov, who was also there, spent much time with Sakharov. It may have been that contact with him was discouraged by the thick air of secrecy about him—I had been hearing since 1953 that he was the creator of the hydrogen bomb.

Speaking Against Nuclear Testing

It was also in 1961 that I learned of an action by Sakharov that attracted publicity. Since 1958 our country had been observing a moratorium on nuclear testing, as had the U.S. In 1961 we were the first to resume testing, offering the obviously contrived excuse that France was said to have been carrying out tests. In fact, this was a raw show of force. By that time Sakharov had developed a most powerful hydrogen bomb, whose explosive power—about 60 megatons—has not since been surpassed. When he learned of the decision to explode the bomb on Novaya Zemlya, Sakharov launched a very vigorous protest against the test.

Sakharov had spoken out against nuclear testing before, but never had his protests been so categorical. He probably felt that these tests imposed a particular responsibility on him. In addition he understood the absurdity of such a nuclear explosion. Such a gigantic explosion was meaningless, even from a military standpoint, since its energy would not have a corresponding destructive effect. The energy would go mostly into the atmosphere, which would then become severely contaminated with radioactive products. From an ecological standpoint, such a powerful explosion would be exceedingly harmful. Sakharov fought to prevent the explosion, arguing his case all the way to the top.

Sakharov was unable to prevent the explosion, but he apparently succeeded in causing the military to start looking at him with suspicion: Here was a man doing some overly independent thinking. His job was viewed as that of simply creating the weapon; beyond that, his advice was not needed—or wanted.

I saw Sakharov engaged in verbal combat in 1964, at a general meeting of the Academy of Sciences of the USSR. At this meeting nominees for academician and corresponding member chosen by the various departments were to be confirmed. The department of general biology had selected Nikolai I. Nuzhdin—one of the closest colleagues of Trofim D. Lysenko—for promotion to academician. The debate over Nuzhdin's candidacy, which drew in Sakharov, Igor E. Tamm, Vladimir A. Engelhardt and Vladimir N. Sukachev, among others, began with some seemingly innocent questions: How many papers had he published? Why had these papers been published for the most part in popular magazines? And so forth. However innocent they may have seemed, these questions and the members' speeches exposed the antiscientific essence of Lysenko's "theories." Particularly sharp and uncompromising was the speech by Sakharov, who directly attacked not only Nuzhdin but also Lysenko. The results of the vote humiliated the Lysenkoites: Nuzhdin received only 14 of about 120 votes.

This was the first time a general meeting of the academy had ever failed to confirm a candidate selected by a department. Because Lysenko was at that time back in good standing with Premier Nikita Khrushchev after a brief fall from grace, these election results provoked intense anger among the authorities. In a report of the meeting that appeared in one newspaper, for example, Sakharov was labeled an "engineer," and it was asked just what an engineer could know about matters of biology.

In 1966 I attended a discussion on "Science of the Future" held by the State Committee on Science and Technology. A Rand Corporation prediction of the future of the U.S. through the year 2025 had been circulated for internal use in the Soviet scientific community, and Vladimir A. Kirillin, chairman of the committee at the time, decided to conduct a discussion of corresponding problems and predictions under the conditions of our own society. He gathered a few scientists, including Sakharov, Nikolai G. Basov, Vitaly L. Ginzburg, Yakov B. Zel'dovich, Bruno M. Pontecorvo, Alexander M. Prokhorov and me. Each participant offered his own predictions regarding the development of the particular scientific fields in which he was working. Sakharov was very interested in the possible use of nuclear explosions for scientific and economic purposes, in particular, to produce immensely strong magnetic fields. Later, in the 1980s, he was to suggest that low-yield underground nuclear explosions be used to prevent earthquakes or to trigger them at a set time.

1968: Final Suppression of the 'Khrushchev Thaw'

The next vivid event linked with Sakharov's name came in 1968. This was a difficult year for all of us. It marked the irrevocable end of the "Khrushchev thaw." In March, student uprisings in Poland were sternly suppressed. On August 21, armed forces of the Warsaw Pact countries marched into Czechoslovakia to put an end to the democratic movement known as the "Prague Spring." That day is remembered by many of us as one of the tragic days of our lives.

That same year Sakharov wrote his famous essay "Reflections on Progress, Peaceful Coexistence and Intellectual Freedom." The treatise was distributed to all the members of the Presidium of the USSR Academy of Sciences. I had an opportunity to read Semenov's personal copy. This essay was a cry of the heart, quite appropriate for the period but coming at absolutely the wrong time in the opinion of the authorities. They fired Sakharov from his job at Yuli B. Khariton's "Installation," a secret atomic research facility. Soon after, Sakharov showed up in Moscow.

Actually, there was a short period in my life during which I *did* see Andrei Dmitrievich fairly frequently. In late 1973, an infarction put me in the academic hospital. Recuperating there at the time were Andrei Dmitrievich; his wife, Elena Georgievna Bonner; and the eminent astrophysicist Iosif Samuilovich Shklovskii. Shklovskii had known Sakharov for a long time, and through him Sakharov and I became better acquainted. Since I was bedridden, they came to visit me, and we had some long conversations. Aleksandr Solzhenitsyn's *Gulag Archipelago* appeared at about that time, and we discussed it.

This account would be quite incomplete without a few kind words about Elena Georgievna. When I first saw her in the hospital with Andrei Dmitrievich, it seemed to me that they complemented each other in a surprisingly harmonious way. When the times were tough, she always supported him, and he her. (When Sakharov and his wife were banished to Gorki in 1980, the corner of 67th Street and 3rd Avenue in New York, by the building housing the Soviet UN mission, was named "Sakharov-Bonner Corner" in protest. Now they will remain there together forever.)

An event that draws a clear picture of Sakharov occurred on February 13, 1974. By that time he had checked out of the hospital, but I was still there. I was sitting with two colleagues in the lobby, reading a plan for a joint paper, when Andrei Dmitrievich appeared. We greeted each other. He said that he had come for his mail, which was still arriving in packets for him at the hospital, several weeks after he had checked out. He said he was happy to see that I was up and around, and then he slipped immediately into what was really on his mind: "You know, an outrageous act has been committed. They have exiled Solzhenitsyn. They have simply thrown him out of the country by force!" Sakharov had come straight to the hospital from the airport; his excited words could be heard by everyone in that overcrowded lobby.

Although I understood and indeed shared Sakharov's feelings, at that particular moment I was somewhat uncomfortable with the idea that his words would drown out the other conversations and would attract everyone's curiosity, which by no means would be benevolent in all cases. I attempted to steer the conversation to another topic, asking him about his health and that of Elena Georgievna. He answered briefly and then lapsed back into the Solzhenitsyn affair; Solzhenitsyn's fate was blocking all other thoughts from Sakharov's mind at that moment. Having blurted out his indignation, Andrei Dmitrievich walked out quickly. My colleagues were curious to know just who this strange man was. Hearing the answer, they forgot about the unread paper and ran out of the hospital to watch Sakharov leave, if only from a distance.

Humanist

I cannot say that Sakharov was the greatest scientist of all time and all places. To say so would be an exaggeration, and Sakharov needs none of that. However, on the basis of an "all-around score"—computed from his personality and his scientific, social and humanitarian activity—Sakharov can legitimately be ranked among the greatest sons of humankind.

We saw Sakharov in three roles: a scientist who created the most destructive weapon ever, a humanist who invested his soul in the effort to see humanity live without weapons in a democratic and free society, and a remarkable thinker, not only in the realm of the natural sciences but also in the realm of the social sciences. We do not know of another such person—at any rate, I do not intend to waste time searching for another example or even for a reasonable approximation.

Looking at Sakharov in his later years, you would see a weak and stooped man, his physical strength apparently on the decline. The poet Yevgeny Yevtushenko has drawn a surprisingly accurate picture by comparing Sakharov's head with a dandelion. Inside this by no means Herculean body, however, thrived an inflexible strength of spirit. Sakharov's strength of spirit was seen particularly clearly during his exile to Gorki, where he went on several hunger strikes, and where his life was in danger (since there are a fair number of active hatemongers in our society).

His inflexible strength of spirit also emerged in words and deeds that at times looked like mistakes to me. I was present at his many speeches at the First and Second Congresses of People's Deputies of the USSR. It sometimes seemed to me that in deliberately stirring up trouble and drawing fire to himself he was making it difficult to reach a compromise solution that might have been a step forward. However, I and many other deputies were apparently not always given to understanding his internal logic or just where his thoughts were leading, so it is difficult for me now, in retrospect, to judge whether one or another of his steps was tactically correct. Indeed, the matter is unimportant. The only important point now is that Sakharov lived and acted among us. I hope that in doing so he succeeded, to some extent, in changing all of us—the entire world—for the better.

Fragments of the Past
(Nikolaĭ Nikolaevich Semenov)

It would seem that the longer and closer the relationship with a person whom one wants to describe, whose image one wants to convey, the easier the job. This is not true at all! I knew Nikolaĭ Nikolaevich for more than 40 years, beginning when I was quite young and essentially ending in old age, and twenty-five of these years were spent in very close and nearly daily contact, both at work and at home, when through routine one involuntarily loses sight of the greatness of someone so close.

How should I tell about all this—long ago past, but always alive in my memory, how should I isolate what is most characteristic and important? Memories crowd together; it is very difficult to arrange them in some kind of order, either chronological or logical, and I have long since given up trying.

Where and when did I first see N.N.? Of course, in the summer of 1942 in Kazan, of course, somewhere in the Institute of Chemical Physics, but I cannot say more definitely. I only remember that after becoming an assistant in the laboratory of Simon Z. Roginskiĭ, I for a long time mistook someone completely unlike him for N.N., and when I found out everyone in the lab made fun of me.

I remember well Nikolaĭ M. Émanuél's defense of his Candidate's dissertation in November 1942, when N.N. very much wanted to raise him immediately to the doctoral level, but did not receive support for this from the scientific council of the ICP. This defense revealed to me one of the fundamental characteristics of N.N.: his democratic nature. When I, on night duty that evening at the ICP, overheard N.N. bitterly complaining to Pyotr L. Kapitsa on the telephone about his (and only his!) bad luck, I sympathized strongly with both him and Émanuél'.

In the same year, 1942, I became friends with the son of Yakov I. Frenkel', Seryozha, who was a laboratory assistant of Boris V. Deryaguin, and Seryozha introduced me to his closest friend, Yuri Semenov, the son of N.N. In the rare hours of freedom from simultaneous work and study at the chemistry department of Kazan University, I visited the Frenkels on Komlev street or the Semenovs at the corner of Dostoevskiĭ and Chekhov streets. We played our favorite card game "hearts" and put on charades.

From this time—the end of 1942 or beginning of 1943—N.N. began to sort of recognize me and from time to time even asked about what I was working on at the laboratory.

In June 1943 I was sent to Moscow, where all the former Leningraders were gradually being transferred: both the colleagues from our laboratory, and the entire ICP. In the fall the Semenov family settled in Moscow, in the Moscow Hotel, room No. 1211, and several of my and Yuri's Kazan-Muscovite friends returned from evacuation, and frequent contacts resumed. Meanwhile, continuing my combined working and studying, I was promoted to level V in the chemistry department of Moscow University, and in June 1944 I successfully defended my diploma study "Catalytic hydrogenation of oleic acid," which was supervised by one of the leading chemical physicists specializing in catalysis, Solomon Yu. Elovich.

The possibilities for continuing to work in the Roginskiĭ laboratory seemed excellent, but an alternative suddenly arose. After the routine conversation with N.N. which at that time had become a sort of repetition of the diploma defense, he suggested that I come directly to him as a post-graduate student. The decision was not easy to make; I was very attached to Elovich, but after thinking the matter over, Elovich himself advised me to take up N.N.'s flattering offer, and in July 1944 I became his post-graduate student.

Shortly afterwards N.N. became immersed in preparing a report to be delivered at the General Assembly of the USSR Academy of Sciences in the fall of 1944. In this report he expressed for the first time the idea that chain reactions are more widespread than even their most ardent proponents had thought, and suggested qualitative criteria for determining which reaction mechanism dominates on the basis of comparing thermal effects (bond strength) and the activation energies of chain growth. I finally got to see N.N. at work with my own eyes, in that state of creative enthusiasm so characteristic of him.

Now I would like to try to define, simply for myself, what were the main creative characteristics of N.N., specific to him, distinguishing his scientific personality from that of our other great contemporaries and compatriots. It seems to me that N.N.'s brilliant talent and his energy and mental power were manifested not in his erudition, his precise knowledge, or in his memory, but in his amazing intuition, the sharpness of his thinking, the breadth of analogies that he came up with, often completely surprising those around him.

It is in just this way that he went, in particular, from the study of dielectric breakdown to a comprehensive theory of combustion and explosion. N.N. was not a "know-it-all;" on the contrary, he rather often upset us with his arguments that it was actually harmful for a scientist to know too much, because this blocks thought and makes it difficult to try a fresh approach to a new problem (because one is afraid to come into contradiction with "firmly established" and "known exactly"!), it produces skepticism too soon and restrains the boldness of the researcher.

If, while working on an article or report in which he set forth new ideas or developed broad generalizations, N.N. needed some data, formulas, numbers, etc., any specific facts, he almost never turned to books or journals, but preferred to directly contact (even in the middle of the night) living sources, experts in the problem of interest.

The birth of a new idea, the appearance of a new problem, a new line of research—whether it came from N.N. himself or initially from outside, such as an urgent need of the country or an order from above—consumed him completely, like a forest fire. During such periods his profound enthusiasm for the new task took N.N. completely outside himself, to absolute, complete concentration of thought and emotions on what appeared to him crucial at that moment.

From his face, his absent-minded, aloof gaze, it was sometimes clear that he was only physically present in conversations on other topics, even when he made remarks; he was actually immersed in his own thoughts, occupied with the intense, unceasing working of his mind. However, he invariably retained a lively interest in everything happening in science, in what was going on at his home institute; he willingly listened to stories of colleagues, went to seminars, and sometimes near the end of a seminar at which he seemed to be only half present he would fluster the speaker with a few questions, and give a completely new emphasis to the entire discussion and to the future course of the work.

Who among N.N.'s closest colleagues does not remember the general calls to emergency jobs, when he had to give a report or a speech, or he was working on a book or a review. Who can forget the many night watches when we young people were literally falling asleep on our feet, while our teacher remained full of energy, good spirits, a desire to listen, think, discuss everything new, smoking cigarette after cigarette, restlessly walking from one end of the study to the other, enveloped in thoughts which we could not grasp, or suddenly running to use the telephone in the middle of the night? And often just when we had become completely stupid and were too exhausted to put together even two words, a joyful exclamation from Nikolaĭ Nikolaevich would ring out: "Well, now it seems I am beginning to understand something." And this modest "something" would turn out to be the key to solving a very complicated problem, the first step toward new ideas and achievements.

During scores of my stay at the ICP I was lucky to get to work with him directly, in close contact, three times; three times I "fell into" the epicenter of the mixture of various scientific interests of N.N., and each time my sojourn in the strong field of his mind enriched me with new ideas, methods, interests, and played an extremely important role in forming my scientific personality. This happened with all of his close students.

The first time this happened was in 1945–1946, when I was still a beginning post-graduate student, and after directing my first efforts toward purely theoretical work (at the suggestion of N.N., I was making a detailed investigation of the double role—inhibiting and catalyzing—of NO in chain reactions of the cracking of organic compounds) I was put under the supervision of Nikolaĭ Mikhaĭlovich Chirkov. The two of us had begun to look for experimental confirmation of N.N.'s hypothesis of the existence of a new type of catalysis—heterogeneous in the place where it occurs, homogeneous in the reaction mechanism—namely, ion catalysis in polymolecular adsorption layers. N.N. would sit for hours in our room, following the readings on the meters measuring the electrical conductivity and pressure of reagent vapors, seeking explanations for everything that seemed strange, making plans for the day after, and the day after that; he sometimes left only after midnight (by that time he already was living in the ICP building), but he would reappear again in the morning. So it went for several months, and all three of us were very happy when N.N.'s expectations were confirmed.

In 1946–1947 nuclear research was begun at the Institute, and in 1949 N.N. had an idea for combining this research with the kinetics of chain reactions in a completely new way. The idea was to ensure intense neutron bombardment of the substance of which the "device" (the customary way of referring to a bomb) is made while squeezing it, thus making it reach supercriticality and explode. Because of the bombardment, the fission chain reaction should begin already at very small supercriticality, and a "clap" ("predetonation")—dispersion of the body of device—would occur instead of a powerful explosion. How could an intense neutron burst be made to occur in a plutonium charge high above the Earth? By means of high-current accelerators producing beams of particles with energies of hundreds of millions of electron-volts, the construction of which also became a part of the program proposed by N.N. and an area of research at the ICP. Work on the nuclear physics part of this program was begun in 1950 at the new Dubna synchrocyclotron by a group of colleagues from our Institute—primarily under the direction of Aleksander A. Koval'skiĭ. I also joined this group, and soon became its head, thereby entering a peculiar school of nuclear physics.

I again crossed paths with the indefatigable N.N., who had become completely immersed in the new idea and the bringing of it to life. On working days he would unexpectedly drive down to Dubna, sit for whole days with us at the accelerator building, and then spend long evenings studying the numbers in our work journals and making estimates. Even on Sundays, when I went to Moscow, it was impossible to get away from work: N.N. immediately began to talk about the experiments and the urgently needed reports.

By the end of 1951 all the reports had been written and sent off, and our part of the work was finished. N.N. soon immersed himself in other problems, gradually returning to "pure" chemical kinetics. At the beginning of 1952 because of the "fight against nepotism" (by that time I had become his son-in-law) I was forced to transfer to the Lebedev Institute, and I was parted from the ICP for nine years. In 1957 Igor' V. Kurchatov allowed me to attend the Gordon Conference in the United States to report on the part of results of studies of the absorption and multiplication of high-energy neutrons. The initial idea of arranging the predetonation of the "device" was at that time completely abandoned and substituted with the idea of "electrobreeding," i.e. of obtaining plutonium due to the spallation and fission in big blocks of uranium bombarded by intense beams of high energy (10^8–10^9 eV) particles. How surprised I was when some American scientists presented reports on nearly the same topic! It turned out that almost at the same time as N.N., similar ideas had been put forward in the United States by the famous Ernest O. Lawrence, the inventor of the cyclotron, and experiments similar to ours, also just declassified, had been carried out at the Berkeley synchrocyclotron. This coincidence clearly showed the high level of N.N.'s ideas, not only in his own field, but also in nuclear physics.

The third example of N.N.'s passionate enthusiasm for science which I witnessed directly dates from 1959–1961, when I started a laboratory of nuclear and radiation chemistry at the ICP and, finally, returned for good to the ICP from the Lebedev Institute. N.N. at this time was immersed in problems of polymer chemistry and polymerization kinetics. In 1958 he essentially became the main driving force behind the convocation of the Plenary Session of the Central Committee of the CPSU devoted to chemistry in general and polymers in particular.

In polymerization kinetics he was mostly occupied with processes occurring in the solid phase, in particular, under the action of radiation. Two laboratories, ours and that of Nikolai S. Enikolopov, carried out long series of experiments. N.N. persistently followed all their details, but refrained from making any conclusions or even predictions. Meanwhile, the International Congress on Polymer Chemistry to be held in the summer of 1960 in Moscow was approaching, and N.N. was supposed to give one of the main lectures.

Literally a few days before this congress, N.N. dropped everything and sat down to work, and we had to respond to the call to arms and form the night watch. But this time I was completely amazed by the fact that N.N., totally unexpectedly, sat down to write the report with a brilliant idea all ready in his head and almost completely polished: the idea of energetic chains of solid-phase polymerization, the propagation of which is similar to exciton transport. His report was perhaps the biggest sensation of the congress, and the subsequent investigations initiated by N.N.'s ideas led to two discoveries: a low-temperature quantum limit of the chemical reaction rate and energetic branching of gas-phase reactions. It was perhaps this time that I was most impressed by the scientific power of N.N. I then tried many times to figure out how he could be so calmly occupied with other things, unworried about the preparation of the forthcoming report; did he have the outline of this report all ready in his head or did he possess an intuitive certainty that when necessary the ideas forming subconsciously would organize themselves and come out right? N.N. never gave any definite answer, but it seems to me that herein lies the secret of his greatness in science.

Nikolaĭ Nikolaevich at home, on vacation, among friends, in general, away from work... . This topic by itself could fill a whole book coauthored by dozens of people. For example, it would be possible to construct a curve from the points corresponding to his jubilees—his 50th, 60th, 70th, etc. birthdays. In my reminiscences about Yakov B. Zel'dovich I told of how we wrote and staged (in April 1946 at the Institute of Physical Problems) the play "Jubilee" for the 50th birthday of our teacher. Various excerpts from this play were later included in many of the social evenings at the Institute, up to the 80th birthday of N.N. and the 50th anniversary of the founding of the ICP (in 1981). The final scene of "Jubilee" was entitled "Forty years later" and showed the celebration of N.N.'s 90th birthday, which in fact turned out to be his last jubilee while he was alive. But now I want to turn to perhaps the most festive of N.N.'s jubilees, his 70th birthday in 1966. One must picture to oneself the huge, packed, well-decorated assembly hall at Moscow State University, the very imposing Presidium on the stage, and in the center the grey-haired Mstislav V. Keldysh presiding, with his slow manner of speaking. Reading through the many congratulatory messages from various notable people, Keldysh announces: "Sergeant Ivanov from the Moscow ORUD-GAI[1] would like to read a message." It should be mentioned here that at this time N.N. was trying to learn to drive a car and obtain his learners' permit. A brave young man in a militiaman's uniform came out onto the stage and read the following greeting:

[1]The Moscow traffic police—*transl.*

To Academician N.N. Semenov
(Address from the Moscow ORUD-GAI)

Esteemed Citizen Academician!

The Moscow ORUD-GAI sends You its greetings and best wishes for Your jubilee.

We would like to assure You that the gentleness and goodwill of our inspectors in their conversations with You regarding minor violations of the driving rules are not due to the Deputy of Supreme Soviet identification card which you showed us by mistake instead of Your driving license, but to Your personal charm. As far as more serious violations are concerned, there have not been any at all, even when, after leaving Your car on Leninskiĭ Prospekt at the request of the ORUD officer, You parked it perpendicular to the sidewalk; the traffic on this main route was delayed no more than ten minutes.

We particularly note Your valuable initiative in showing gratitude for the technical assistance, expressed in the fact that You paid not only the driver of the taxi which towed You, but also the passengers. At the present time a strict selection is being made from among the thousands of taxi passengers wishing to enroll in special groups which will go to make a living in the area where the Institute entrusted to You is located.

However, we must conclude our greetings to You by saying to You, Citizen Academician, that let us not pretend that You are seventy, it is really ridiculous, no one would believe You; that is to say, we would treat Your driving violations leniently anyway. However, such violations are already nearly unobservable. Your driving is getting better all the time, and in the very near future You will undoubtedly become one of the best amateur drivers among the vice presidents of the USSR Academy of Sciences, the headquarters of Soviet science.

For the head of the Moscow ORUD-GAI

Colonel A. B. Avtodorov[2]

Aside from the author of the letter (the present author) and Keldysh, only a few people were let in on the joke beforehand, and only a few recognized Sergeant Ivanov to be Mikhail B. Bulygin, a colleague at the ICP. Therefore, the greetings from the ORUD-GAI caused complete shock, but the most surprising thing was that the recipient took it seriously.

For many decades the favorite pastime of N.N. was hunting, and his most frequent companions for this were Academician Victor N. Kondrat'ev and Eugen K. Russiyan, head of ICP workshops. At home he always had a hunting dog, usually a red Irish setter: Ralph, Toĭ, Alpha, Sil'va, and, finally, Alan, presented on a pillow to N.N. for his 70th birthday. For the same birthday Anatoly P. Aleksandrov presented N.N. with a shotgun with a curved muzzle for shooting from an angle. Incidentally, later it was necessary to

[2]Avtodor was the name of the Soviet society of professional and amateur drivers.

register this shotgun on the grounds that its end could be sawed off. It may be that N.N. sometimes had some particularly successful hunting expeditions—but I cannot recall any. However, he would return from the hunt empty-handed but full of impressions, good cheer, and a desire to immediately return to work. For many years I have recalled with a smile his enthusiastic story about one hunt with a very modest catch: "The ducks were flying around like mosquitoes."

Another characteristic of N.N. was his ability to instantly stop and take a rest, to give himself a respite before attacking a new, difficult task. If a free window of 10 minutes suddenly appeared, he could sleep for exactly those 10 minutes and wake up refreshed. And if a free hour or so arose, he loved to sit down to cards and play the game "podkidnoĭ durak"—with passion and even a kind of scientific approach, working out what the different possible endings could have been for a given final set of cards on the table.

We spent many summers together, most often at the dacha in Lutsino near Zvenigorod, and several times far from home. Starting from the first year the dacha settlement was built (1948) in Lutsino, many Academician-chemists lived there, including N.N.'s closest friend, Aleksander N. Frumkin, Pyotr A. Rehbinder, Aleksander N. Nesmeyanov, and Aleksander N. Terenin (and later also Victor N. Kondrat'ev, Victor I. Spitsyn, Ivan L. Knunyants, and Nikolaĭ M. Émanuél'). It was traditional for him to take long walks with Frumkin accompanied by two dogs, to celebrate July namedays with the Rehbinders, and sometimes to play skittles with Nesmeyanov and his guests. I remember how once N.N. urgently needed to discuss some scientific problem with Terenin and he asked me to go find him. It was broad daylight, and for some reason N.N. became very gleeful when I returned and said that Terenin was sleeping. He jumped up from the table and, inviting me to follow him, ran straight to Terenin's dacha in pajamas shouting "Let's go wake up Terenin." Terenin, of course, could not withstand this onslaught and, shaking off the last bit of sleep, obediently followed N.N. to his dacha. When there were still some places with unspoiled nature around N.N. loved to go mushroom hunting.

We often played croquet with great enthusiasm. A period of passion for canoeing came and went; the least pleasant aspect of it was the nearly daily need to hoist the canoe up on our shoulders and carry it from the river up a long hill and over a fence surrounding the area. Once the three of us (including my wife Mila Semenova, N.N.'s daughter) went by canoe down the Moscow river to Nikolina Gora, where Kapitsa, another very close friend of N.N.'s dating from student days, had his dacha. N.N.'s chauffeur Aleksei D. Gur'ev met us there with the car; his family also spent the summer in Lutsino, in a small house/garage near the dacha.

It was always interesting to be present at conversations between Semenov and Kapitsa; it is a terrible shame that at the time we were not able to record them. The topics were wide-ranging: science, academic news, household business. It was funny to see how Kapitsa, who very much liked spicy anecdotes, made N.N. turn red listening to them. From the mouth of N.N. one could hear (but only very rarely, after he began to socialize with generals and bomb makers) only a few unseemly variants of the word "khrenovina" ("khren" means horseradish).

As far as I remember, of the very large circle of N.N.'s acquaintances only a few were his close friends, with whom he was on a familiar footing: the Kapitsas (Pet'ka and Anechka); Frumkin (Frumochka); Frenkel' (Yasha or Yashen'ka); expert in physiology of vision Vera G. Samsonova (Verochka), the wife of physicist Iosif L. Zel'manov; that perhaps is all. N.N. called very few people by name; even scientists with whom he worked closely like Kondrat'ev, Khariton, Émanuél', or a close friend like Sergei S. Medvedev usually were called by name and patronymic. However, this did not imply that it was difficult for N.N. to become close to people. On the contrary, in any social group—be it the young people at the Institute and in the department or people resting at a sanatorium, at scientific meetings or on hunting or fishing trips—N.N. did not feel himself to be an "immortal," but just a simple man who deeply loved and was able to have a good time.

Nikolaĭ Nikolaevich was a star at the Institute parties, even when he was well over 60, owing to his surprising mastery and grace in dancing the mazurka; he delighted everyone present by his miraculous transformation into a handsome young man who seemed to belong to balls known to us only from movies and literature.

N.N. had the makings of an actor, as was manifested in charades, our traditional amusement both in Moscow and on vacation. At Koktebel in Crimea, which we all loved, our movie stars Sergeĭ Gerasimov and Tamara Makarova became the directors, and theater producer Boris Zakhava assumed this role when we spent summer at the journalists' rest home near Varna. In the Zakhava production N.N. was instructed to growl like a lion.

Although N.N. could not boast of having a musical ear, he loved to sing in chorus among friends. Among his favorite songs were "From the island down the deep stream" and "The spinner," and later he very much liked Vladimir Kandelaki's rendition of "Come to us, genatsvale" (i.e., friends in Georgian).

Natal'ya Nikolaevna (also N.N.!) was N.N.'s guide in classical music, and their daughter then followed in her footsteps. N.N. (he) and N.N. (she) used to go to the conservatory, and for her sake he fairly often attended parties at the institute's kindergarten, which always enjoyed his patronage. N.N. (she) was an invaluable companion and helper to N.N. on trips abroad: he could not speak in any foreign language, but she knew three: English, French, and German.

However, "N.N. abroad" is another topic, and I think it will be written on by many others (in particular, I have already been able to read the very lively story of Academician Nikolai A. Platé—then quite young Kolya Platé, with whom both Semenovs, the present author, and about 20 other chemists from our institute attended the IUPAC Congress in Canada in 1961).

As an example of a trip taken with N.N. I would also like to tell about the summer of 1951.

However, first I must make a "lyrical digression," or else it will not be possible to understand the composition of our group that summer. At the end of 1948 some so-called "secretaries" were assigned to N.N.; in fact, they were bodyguards from the KGB, then assigned to all the leading scientists working on atomic research. These secretaries took turns being on duty 24 hours a day, slept in a tiny room in N.N.'s apartment, and relentlessly followed him everywhere. First there were two, and by 1949 a third had been added. The people who were put under such supervision displayed a wide variety of reactions to the appearances of the secretaries. Landau threatened suicide. Zel'dovich would not let his secretaries into his apartment and they slept on the staircase. As a result, the authorities were forced to retreat in these two cases. N.N. reacted completely differently. He had a very sociable nature, and accepted the positive aspects of having secretaries. In fact, the perpetual problem of looking for a fourth partner for games of "podkidnoĭ durak" vanished, and he acquired new companions for hunting and various other expeditions (like the trips to Dubna mentioned above). Moreover, they could always remind N.N. when and where he had to go, keep a supply of cigarettes and matches, help him find his constantly disappearing glasses ("Here, Nikolaĭ Nikolaevich, they are in your shirt pocket"), etc. etc. One of N.N.'s three secretaries, Pavel Semenovich Kostikov, later became genuinely close to our entire family, an irreplaceable assistant to N.N. both at the institute and in the Presidium, and actively took all N.N.'s interests to heart. In the first year or two of his position as secretary Kostikov one summer in Lutsino uttered a famous line which we recalled with laughter many years later. At that time N.N. had a black dog named Sil'va, the most stupid of all in his series of hunting dogs. After returning from accompanying N.N. on a walk in the woods with Sil'va along, Pavel Semenovich sincerely declared: "It's really a disaster when this Sil'va comes along. You just don't know who of the two (N.N. or Sil'va) should be shadowed".

So, in the summer of 1951 there were six of us: elder Semenovs, Mila and I, Pavel S. Kostikov, and the "most senior" of the three secretaries, Ivan Timofeevich Kurinnoĭ, a pert and saucy fellow who was also sincerely attached to N.N.

After arriving in Odessa and spending a couple of days in that wonderful and unique city (at that time some of the typical dialect of Odessa still remained in the speech; I remember how we enjoyed overhearing on the street the words "We need to find out just how he drank up so much of her blood"), we settled down in the "Russia" and set off for Batumi. Lying in the sun on the deck, I and N.N., Kurinnoĭ, and a varying fourth partner often played "podkidnoĭ durak." In Batumi it was extremely hot and also very humid. The elder Semenovs and Mila and I stayed all together in a very large deluxe room and were simply exhausted from the heat. At night we would fill the bathtub with water, and when it was impossible to sleep each person went into the bathroom, immersed himself in the water, and tried again to sleep before the water evaporated from his body. I remember how we all with N.N. went to the cinema and watched with great satisfaction an old movie which had been famous before the war and had returned to the screen for a short time: "Peter" with Francesca Gaal' in the leading role. The climax of our trip to Batumi was a sumptuous banquet held in honor of N.N. by Victor Ivanovich Kokochashvili at his estate (a dacha with very large grounds) in Makhindzhauri. The eminent Georgian scientist Kokochashvili worked at the ICP in the 1930s in Leningrad and literally worshipped his teachers Semenov and Zel'dovich. At the end of the 1940s he got married at a somewhat advanced age. Charming and deeply intelligent (a match for her husband), Nastasiya Davydovna was the hostess of the party and Victor Ivanovich was the toastmaster.

Before our trip to Makhindzhauri there was a small fight between N.N. and the senior secretary Kurinnoĭ. The latter insisted that both secretaries attend the banquet, because it was a risky affair. N.N. got really infuriated: "When I wanted to go boar hunting neither of you wanted to go, but when a banquet comes up you both do," but he soon cooled down and yielded to the secretary. I will not begin to describe the banquet itself, the huge meal in the garden with nearly a hundred guests, numerous toasts (beginning, of course, with toasts in honor of the genial leader and luminary of all the sciences—Joseph Stalin). I will only say that the moved N.N. himself offered many toasts in honor of the host and hostess, to the glory of science, and to the prosperity of Georgia. We returned to the hotel late at night, and on the following evening went to Sukhumi on the small steamboat "L'vov." There was a great storm and the boat was rolling around; we all were lying there nearly dead except for N.N., who demonstrated an amazing resistence to seasickness. Not without reason had his students and friends in 1924 in making up a song in the Crimea about Batiliman nicknamed N.N. their captain!

Our trip ended in Novyĭ Afon with a nearly month-long stay at the Abkhazia sanatorium. N.N. gave himself a real rest, and it will be seen from the following section of these reminiscences that this was a very difficult time for him and he very much needed a rest. It was affecting his enviable ability to temporarily put all unpleasant thoughts out of his mind and to think only of the present, and to completely relax. We roamed around the area and spent a couple of days at Lake Ritsa, where to N.N.'s great joy we met Sergeĭ Sergeevich and Ol'ga Mikhaĭlovna Medvedev. In Novyĭ Afon itself on a little island in the middle of a pond was a small, cozy restaurant which served excellent roast quail. N.N. liked to go there often and over food and wine tell about his past successes quail hunting in the Crimea. The future Academician Vitaly L. Ginzburg and his wife Nina were staying in the same sanatorium as us, so our group on the beach grew. At that time the train through Novyĭ Afon seemed to pass right through our sanatorium, and Ginzburg would say: "This is the first time I am staying in a sanatorium where a person's voice cannot be heard over the trains." Near the end of our stay in Novyĭ Afon, Vladislav V. Voevodskiĭ and a group of colleagues from the ICP appeared, on their way to the sea after coming down from a mountain expedition. N.N. was on very good terms with Voevodskiĭ, and in an unofficial setting often called him Vladik. Vladik's appearance was most welcome to N.N., and was a sort of warmup for his return to work in Moscow. They conversed enthusiastically on the beach about various kinetic problems; it seems to me that it was at that time that the idea was born of a possible role of chain reactions on a surface in heterogeneous catalysis, which was later the subject of a well-known study by N.N., Voevodskiĭ, and Fyodor F. Vol'kenshteĭn. It was decided that N.N. had had a good rest and was, like the well-known sports motto, "fit for work and defense." He would have to fight back against the furious attacks of Akulov and his clique, it was urgently necessary to continue his nuclear research, and, in addition, a crop of new ideas was insistently ripening inside his head and drawing N.N. to his favorite topic, chemical kinetics. For me the forthcoming year saw the beginning of my nearly 10-year separation from the ICP.

There are several curious incidents which I heard about many times from N.N.; he himself loved to talk about them in various groups of people, and so I decided to include them in these recollections.

At one of the banquets at the Academy of Sciences, N.N., going up to high-ranking playwright Academician Aleksander E. Korneĭchuk, turned to him with the words, "No, Korneĭchuk, you are not Shakespeare." Korneĭchuk was completely astounded and only much later approached N.N. in turn and said: "And you are not Newton, Nikolaĭ Nikolaevich." N.N. burst out laughing: "Correct, but you should have answered immediately, not an hour later."

In the years before the war the singer Deborah Ya. Pantofel'-Nechetskaya, a coloratura soprano, became widely popular throughout the country. Sincerely wanting to praise her, N.N. said: "It is simply amazing how you, with such a tiny voice, can sing so remarkably," and was surprised that the singer was not happy with this compliment. Famous Bass Mark O. Reĭzen reacted similarly when, also with good intentions, N.N. praised him with the statement "We do not have a second Chaliapin, but you are slightly reminiscent of him."

During the first years of the existence of the ICP in Moscow a magnificent rose garden was planted in a large bed in front of the institute building. One beautiful day an unknown man entered N.N.'s office and said, "I am Tsfasman. I came to ask you for a few rose bushes." The puzzled N.N. decided that the visitor must belong to some important regional authority. Without entering into conversation with the visitor, he handed him over to his deputy building manager, instructing him to give Comrade Tsfasman whatever he asked for. Only that evening, at home, did he learn that he had missed the opportunity to become acquainted with a famous jazz musician.

When we read about the many awards and honors given to N.N. by our country and the entire world, it can appear that his entire path through life was cloudless, strewn with only roses and no thorns. It was not that way. For more than 10 years he underwent downright persecution from a group of physicists and chemists at Moscow State University who, in that shameful affair, clustered around one of the most odious figures in the physics department at that time, Nikolai S. Akulov.

Akulov was a physicist who had done some good work on magnetism, but who clearly was far from reaching the level of our leading scientists. He well understood this and was particularly envious and aggressive. Not long before the war, Akulov suddenly decided to work on the theory of chain reactions—the lifelong occupation of Semenov and his school—and published a small book on this topic suggesting a new theory. Welcome! It could only be greeted in this way, and Akulov was invited to Leningrad to give a seminar at the ICP. But there was a problem: it turned out that all the innovations reduced to elementary exercises in mathematics, aside from the fact that the most general features of chemical chain reactions were distorted. In the end the new "theory" turned out to be purely formal, with no relation at all to chemistry and in contradiction with a large set of facts. This was explained honestly and directly to Akulov at the ICP.

After a three-year hiatus related to the evacuation, the dialog with Akulov was resumed in Moscow in 1943. By this time Akulov's booklet was beginning to be widely publicized as the catechism for studying chain reactions by a group of his companions-in-arms (whose work in turn was exalted by Akulov in every way possible), and accusations of Machism were made against N.N.

N.N. with his extreme conscientiousness simply could not understand that scientific arguments are useless when confronted with base motives, and at first he seriously thought that Akulov had simply not completely understood something, that if his error was explained to him the fight would be over. I remember a three-hour telephone conversation in which N.N. literally exhausted himself trying to make Akulov understand his errors. Hanging up the phone, N.N. joyfully exclaimed: "Well, now it seems he finally understands."

A vain hope! This was followed by a torrent of accusations against N.N.: accusations now not only of philosophical errors, but also of toadying to the West. This latter accusation was made because Semenov had dedicated his famous book *Chain Reactions*, for which he later won the Nobel Prize, to his "Distant teachers," the Swede Svante Arrhenius and the Dutchman Jacobus van't Hoff.

In 1946 the falling into disfavor of Kapitsa, who was essentially under house arrest for many years at Nikolina Gora, gave new strength to Akulov and his closely knit group of professors in the physics and chemistry departments at MSU. Following the well known principle of "stop thief!" the Akulov group began accusing Semenov and Kapitsa of cliquishness, and, to be more convincing, adding to them their common teacher Abram F. Ioffe, Frumkin, and others of our greatest scientists.

Akulov attempted to convince everyone that Semenov was not the creator of the theory of chain reactions. I remember a funny instance, when a student was egged on to ask Semenov, at the next regular discussion, the question of what exactly was the difference between the Semenov theory and that of the Danish scientist Jens Christiansen and the Dutch scientist Hendrik Kramers, who also had worked on chain reactions. The student punctually got up and began to ask the question. However, he forgot the surnames Christiansen and Kramers, faltered, and stood on the platform without finishing the phrase. After smiling, N.N. bitterly asked him, "Probably you have Christiansen and Kramers in mind?" The student became very embarrassed.

However, in a short time, since the political situation in the country was changing, Christiansen and Kramers were thrown out, and it became necessary to find some native expert to substitute for Semenov. And one was "found": a good Russian chemist, who had successfully worked on both adsorption and on conjugated chemical reactions (but not at all on chain reactions!), Nikolaĭ Aleksandrovich Shilov, who had died in 1931. In fact, Shilov had known Semenov's work and valued it highly; he never in his life pretended to be an expert on chain reactions.

During those years N.N.'s friendship with Kapitsa (in spite of the fact that N.N. was forced to completely break off contact with him in 1946; only Natal'ya Nikolaevna continued to travel regularly to Nikolina Gora) cost him yet another humiliation, perhaps the deepest of his life, of which nothing has been said up to now.

When in 1946 the ICP was drawn into research on the atomic problem, the most important task assigned to the Institute was connected to building the test site under Semipalatinsk (today widely and, one can say, sadly known to the entire country and the entire world) and making various possible measurements during the testing of the first Soviet atomic bomb, to be exploded within a few years. The future Academician Mikhail A. Sadovskiĭ became the closest assistant of N.N. in this work and the immediate supervisor of that problem.

At the beginning of 1949 N.N. made the long trip, difficult in winter, to carry out a detailed, fault-finding examination of the progress of the preparations under Semipalatinsk. Of course, the names of that town and other nearby places were never mentioned at that time; they were top-secret, and the made-up term "land of Limonia" was used at the institute. Summer arrived, and dozens of colleagues from the ICP were sent to Limonia. N.N. for some reason remained in Moscow. The bomb was exploded on August 29, the people began to return home; I sensed from some nervousness in N.N., that something was wrong, but, of course, I did not speak with him about secret topics with which I was not directly involved. However, the cat got out of the bag, and I soon learned from our "Limonians" that N.N. had simply not been allowed to go to the test site. Moreover, the colleagues from the ICP who had gone to Limonia were summoned to the Special Section and warned not to say anything about the tests to their director after their return. Knowing now of the hatred Beria had for Kapitsa, I have no doubt that the disgraceful discrimination against N.N. in 1949 also came from him.

The many-year-long discussion with Akulov took a considerable amount of energy from N.N. and his scientists, distracted them from incomparably more important work, and, in addition, proved completely fruitless, as there were no scientific arguments involved at all, and in principle none could be involved; after all, science as such was the subject of least interest to Semenov's opponents.

In the spring of 1952 the Akulov group organized a two-day mass meeting in the MSU physics department. The presentation of the great mathematician Andrei N. Kolmogorov, who defended the Semenov theory and pointed out the errors in the Akulov formulas, and the beautiful arguments of perhaps the most talented young representative of the Semenov school, future Academician Voevodskiĭ, were contrasted with a magic lantern: a demonstration was given of how many times foreign scientists were cited in the articles of Semenov and his colleagues. This example of a "scientific controversy" in the Stalin era was presented without restraint by the Akulov group to an auditorium full of students.

One of the two days of the "discussion" at MSU coincided with N.N.'s birthday. We and a group of N. N's closest colleagues went directly from the university to his house, and there Kondrat'ev read his very artistic verses addressed to the "birthday boy." I give them here.

FROM VICTOR N. KONDRAT'EV TO NIKOLAĬ N. SEMENOV
On his birthday, 16 April, 1952

The director is baking the directives
And scribbling scientific papers.
But there already the willows have arrayed themselves
For a holiday of sun and water.

There the rivers have overflowed, like the sea,
Into the forests, into the fields,
And the alarmed cry of the crane
Is lost in the great blue space.

There the breast lightly and freely rises and falls
And one smells the harsh odor of rotting leaves.
There at daybreak from afar can be heard
The soft beating of ducks' wings.

In anticipation of the springtime battle
The little black ducks scurry around,
I look at the old, crumpled map:
All directions are good!

In the distance a springtime song rings out,
And there is no strength to resist it...
There are no such immense or wonderful spaces
Anywhere but in my homeland!

May you always be blessed!
I am proud of you, I sing of you,
You are the only one in the Universe,
My dearest Russia!

... The simple hunting equipment:
The shotgun... Do not stint on the cartridges!
Nastya has already prepared the food,
And Gur'ev awaits you at the door.

Now we can be on our way! Away problems
And neverending work,
Reports, reviews, accounts,
No end to them, impossible to count them.

How quickly the miles have passed,
We have crossed that sweet boundary
Beyond which lie sunny vistas
Veiled in a blue haze.

Here in the sound of the drifting ice
We suddenly realize, yet again,
What marvels Nature conceals
For each of us!

Here in sweet drowsiness
Under the quietly whispering reeds,
In the cozy little blind
One can forget about ducks, about hunting.

And when the busy day is over,
Before the new morning arrives
To live by this amazing life
At invigorating camp-fire.

Tomorrow, rising at dawn,
You might hear a thrilling "bang!"
And wake up the morning with a strike...
All this is possible, but impossible!

After all, tomorrow at the university
Another Akulov's witches sabbath
Another report at the technical council
And the usual daily muddle.

Things pass by and disappear,
Like snow, like the spring floods,
Weeks, months, years
Flow under the mare's tail.

Not for you are the willows among fleecy clouds
Or the sound of the rebellious water!
... The director is baking the directives
And scribbling scientific papers...

As expected, there were administrative consequences of the meeting at MSU; in September, 1952. Voevodskiĭ was dismissed from the MSU chemistry faculty.

However, the Stalin era was coming to an end. 1953 began.... The new age led to new criteria for judging the worth of scientists: their talent and their achievements in science began to play an ever greater role, and it became more and more difficult to get anywhere by demagoguery and denunciations. Kapitsa returned to his institute. Under these conditions the activity of Akulov and his group began to fade away quite rapidly. Akulov himself had been elected to the Belorussian Academy of Sciences back in 1940, but for 15 years had only received his wages from it, as he worked far from Belorussia; he was unwillingly forced to return to Minsk, and the physics department of MSU abandoned a few of his colleagues. A new dean of the department was appointed. Many eminent physicists entered the university, and even the huge lecture halls of the new MSU building were hardly big enough to hold all those attending their lectures.

In 1956 Nikolaĭ Nikolaevich became the first (and in chemistry the only) Soviet winner of the Nobel Prize, and this event, which became a genuine celebration of Soviet science, fundamentally changed the relationship between N.N. and the authorities.

N.N. was soon elected academician-secretary of the Division of Chemical Sciences, and in the spring of 1958, as already mentioned, his persistence and energy played a crucial role in preparing the Plenary Session of the Central Committee of the CPSU for the development of chemistry. The same year brought joyful events for the Semenov school: three of its scientists, Émanuél' (for Moscow) and Voevodskiĭ and Koval'skiĭ (for the Siberian Branch) were elected Corresponding Members of the USSR Academy of Sciences, and Émanuél' also became a Lenin Prize winner.

After the reorganization of the USSR Academy of Sciences in 1963 N.N. was elected vice president of the Academy of Sciences and chairman of the Section of Chemical and Biological Sciences.

Therefore, during the 1960s N.N. was universally recognized not only as the ideological leader but also the administrative head of Soviet chemistry. During those years he also played a very important role in the revitalization of Soviet biology and fostered the development of molecular biology and molecular genetics in our country. It is impossible to overestimate the importance of his remarkable writings revealing the full extent of the damage wrought by Lysenko and his group.

One of the main demands which N.N. made for the successful development of Soviet science was its rejuvenation. To achieve this, he set about replacing directors of chemical institutes over the age of 70. I involuntarily overheard a series of telephone conversations which resulted in the replacement of the directors of the N. S. Kurnakov Institute of General and Inorganic Chemistry (IGIC) and the Institute of Organic Chemistry (IOC): Ilya I. Chernyaev was replaced by Nikolai M. Zhavoronkov and Boris A. Kazanskiĭ was replaced by Nikolai K. Kochetkov.

Beginning in 1964, N.N. many times spoke to me about the idea of turning the direction of the ICP over to Émanuél'. Because of this he was particularly pained by the failure of Émanuél' to be elected Academician in 1964. He was lacking a single vote, and N.N. drove to visit Chernyaev at his dacha and literally begged him to vote for Émanuél'. It wasn't very surprising that Chernyaev, who had just been replaced as head of the IGIC, flatly refused N.N.'s request. A happy outcome of these elections for all of us was the election of Voevodskiĭ as Academician.

In the summer of 1996 N.N.'s stay in the dacha at Lutsino was also filled with academic duties. He went around to all the chemist-Academicians in connection with the forthcoming elections and this time achieved success: Émanuél' became an Academician. In a conversation with me, N.N. confirmed his idea of him as his future successor at the institute. He also began to think of retiring from the post of vice president in the elections coming up in 1967, but it turned out that Keldysh did not support the candidacy of Émanuél'. During those years Voevodskiĭ was renting one of the dachas at Lutsino, and one beautiful day N.N. called me to come with him to visit Voevodskiĭ for a very serious discussion. After some small talk, N.N. directly suggested to

Voevodskiĭ that he prepare to return to Moscow within a year as vice president of the Academy of Sciences. After two more meetings with N.N., Voevodskiĭ finally agreed. Alas, fate willed differently: on February 20, 1967 Voevodskiĭ died, five months before his 50th birthday. The talk of N.N. retiring from the vice presidency stopped, although Keldysh's opinion of Émanuél' clearly began to go up at the same time. The urgency of the question of transferring Émanuél' to the directorate of the ICP also visibly faded away.

Meanwhile, the idea of bringing younger people into Soviet science won new supporters, and this became more and more apparent as the elections for the directorate of the Academy in 1971 approached.

The previous events of 1970 were in many respects distressing for us. In the winter a general meeting of the Academy was held, and at it all the vice presidents reported on the research trends in their fields. It must be admitted that oratorical skill was not one of N.N.'s strong points, and he had a special problem with the time limit. These faults were quite apparent in N.N.'s report on chemistry at this meeting. An enormous amount of material was presented, without any indication of what was more and what was less important and without any classification; instead of the allotted 40 minutes, the report took two hours; it was difficult to hear, the noise kept increasing, Keldysh became nervous and angry, and finally, after several warnings, stopped N.N.'s presentation. Knowing how much time and effort the preparation of this report had cost him (and indeed all of us) and how important it was, we left the meeting in gloomy spirits, not looking at each other, full of bad premonitions. In fact, after a few days we found out that right after the meeting one of our chemist-Academicians had gone up to Keldysh and asked that the chemistry head be replaced and met with a sympathetic response. Two months later the Committee on the Lenin and State Prizes of the USSR in science and technology refused to award N.N. the Lenin prize for his remarkable book *Some Problems in Chemical Kinetics and Reactivity* (he was awarded this prize in 1976).

In 1971 Andrei N. Belozerskiĭ replaced N.N. as vice president of the USSR Academy of Sciences. N.N. turned 75 the same year. The day of his jubilee began for me with the funeral of a remarkable man and scientist, Igor' Evgen'evich Tamm. During this funeral I learned from Tata Khariton (daughter of Yuli B. Khariton) of an important event in the life of N.N. (his departure from the family), which made future meetings between us much rarer. I therefore break off these reminiscences and stories about N.N. in 1971.

Recalling the past is also a sort of chain reaction. One set of recollections leads to many others, and stopping the flow is not really easier than writing the first lines. Therefore, here I will simply stop my tale of the past, but first I would like to briefly discuss the remarkable human features of N.N., which seem to be so rare nowadays: his sympathy for the problems of others, his caring about people, and his lack of self-interest.

One of N.N.'s favorite maxims to young scientific colleagues was "the worse a man lives, the better he works." At first this was very intimidating—after all, life for someone starting out in science is difficult, and it seemed that it was not possible to count on the sympathy and aid of N.N. in view of such statements. However, it only appeared this way. Actually, probably none of our academic directors worried about his coworkers as much as N.N. After the institute was moved from Kazan to Moscow, it was first of all necessary to find places for people to live; after all, the entire old guard of the institute consisted of Leningraders who were essentially homeless in the capital. And here a real miracle took place: only eight months after the end of the war, people were beginning to move into the first new houses (made of wood and now already demolished) on the ICP grounds. Others were built, some our own and some obtained from the Academy of Sciences and other departments. Incidentally, both in Leningrad before the war and in Moscow N.N. let colleagues from the institute live even in his own apartment for quite a long time. In 1945 N.N. dressed the ICP workers from head to foot, obtaining from wherever possible (including American donations through the Lend-Lease program) overcoats, suits, and shoes. Every worker who went to N.N. to ask for medical help for himself or his family came away feeling a dependable hand of support. I remember this both from my own experience (in the Fall of 1945 I had paresis, paralysis of a facial nerve, and N.N. with difficulty arranged for me to receive a course of physical therapy at Uzkoe) and from that of Émanuél', who fell ill in 1946 with tuberculosis and with the aid of N.N., who was very worried about him, obtained treatment at a hospital and a sanatorium. There are also many other examples.

Later on, as a member of the family, I was amazed at how much N.N. worried not only about more or less serious illnesses of people close to him, but about anyone who was late in arriving at an appointed time. He always feared that something terrible had happened. For example, he often went around to all the neighbors himself looking for our eldest son Mitya, whom he particularly cared about, whereas we were not alarmed at all and were sure that Mitya had simply gone off with someone and forgotten the time. In the summer of 1966 a group of 10 of us went to Yugoslavia. Once I walked to the cinema, and on my way back missed meeting up with N.N. at an appointed spot. After waiting for a while for him in vain, I returned home and found everyone alarmed by N.N.—he had already organized a search party for me.

On Sunday evening, January 7, 1962, we were playing cards when the telephone rang. Shal'nikov informed N.N. about the automobile accident in which Landau was injured, and told him that pure urea preparations were urgently needed. Extremely upset, N.N. phoned up scores of organic chemists and after several hours had reached everyone accessible in Moscow. Meanwhile, the first such preparations, specifically for medical use, were rushed from London.

After becoming vice president of the USSR Academy of Sciences in 1963, N.N. in this post displayed a completely unique disinterestedness. A person in a high position usually "pulls the blanket over himself," and, alas, we know of many examples of such people who found institutes for the next century which eat up the budgets of others in this century. On the contrary, N.N. as vice president often refused his own colleagues completely legitimate requests for foreign currency, funds, and equipment so that no one could accuse him of favoring his own institute at another's expense.

Things did not always go easily for him. In particular, there was the case of the nomination of Frumkin to receive the award Hero of Socialist Labor on the occasion of his 70th birthday in 1965. N.N. followed every step in the progress of the paperwork in a very hostile atmosphere. On the day of the jubilee for Frumkin he arrived at the banquet rather late, but with good reason. Triumphantly rising from the table together with Frumkin, he read the just-signed announcement of the high award and kissed Frumkin, radiant with joy for his friend.

We got used to the fact that the joys and sorrows of his loved ones, friends, and colleagues were for N.N. the same as his own, and accepted this as something intrinsic to and unchanging in him. It was natural to respond to him in the same way—with a combination of esteem, admiration, and love. That is how it was, that is how it remains in our memory, and that is how we must pass it on to our descendants.

After writing these fragmentary reminiscences I read the text written by Pavel S. Kostikov, and found one clear contradiction between what I and Kostikov remember.

I wrote that after 1946 N.N. was forced to stop visiting Kapitsa at Nikolina Gora. Kostikov remembers such a trip being taken the first day he worked for N.N., July 9, 1949, when Pyotr Leonidovich was 55. My efforts to find a court of arbitration in the persons of Natal'ya Nikolaevna Semenova, Anna Alekseevna Kapitsa, and the eldest Kapitsa son Sergeĭ were in vain. None of them remembers N.N. visiting Nikolina Gora in July 1949, and Sergeĭ added that N.N. attended his wedding in Moscow on October 11, 1949 and there met his father.

However, it is difficult to imagine that Kostikov could be mistaken about that first, memorable day of his acquaintance with N.N.

I would not have dealt with this not terribly important discrepancy in details of the distant past if it had not illustrated so clearly a possible direct causal relationship between a trip (almost certainly the only one) made by N.N. to the dacha of his oldest friend, fallen from favor, and the removal of N.N. from our first nuclear test, described above, in which Beria probably played a role.

I also asked Academician Sadovskiĭ, one of N.N.'s closest friends, companions-in-arms, and disciples, about that time. He told of how he was summoned to the science division of the Central Committee of the CPSU(B), where the then-director of that division pressed him for information about the political and ideological views of N.N., asked how strong his patriotism was, and sought to compromise him on the basis of his worldwide fame as a scientist. Sadovskiĭ indignantly rejected all such questions, and they gave up on him with ill-disguised ill will.

Finally, there is another fact for the historical record, which has just been communicated to me by Anna Alekseevna Kapitsa. Hundreds of thousands of people now delight in the recently published correspondence of Kapitsa with Stalin, Molotov, and other leaders great and small. These letters, which were kept in the Kapitsa dacha at Nikolina Gora, could at any moment have been removed in a search, and the world would have irretrievably lost this invaluable evidence of the inflexible spirit and moral character of Kapitsa as well as of the deepest scorn of the light of the human mind by tyranny and its associates. To keep these valuable documents from harm, Anna A. Kapitsa entrusted them to Natal'ya N. Semenova, her oldest friend, dating from high school days, and all the letters were kept for eight years in the Semenov house, unbeknown to either N.N. or Kapitsa.

My Remarkable Relative Shura
(Aleksander Iosifovich Shal'nikov)

He usually appeared without warning, completely unexpectedly. Our apartment was one of his stops in his circuit of the apartment house of the Institute of Chemical Physics on Vorob'jevskoje Shausse, 2 bis, where his old friends Ryabinins lived for a long time on the first floor, other close friends- Leĭpunskiĭs on the second, and we on the third. After the usual rebukes (instead of greetings) that we were swine for not phoning him or visiting, Shura would take off his jacket and come with me to my study or the kitchen. He almost never sat still; during every one of our conversations in my study (which usually included my wife Mila, Lyudmila Nikolaevna Semenova) he leafed through my books, and during conversations in the kitchen he engaged in his favorite activities: preparing (very good!) tea or washing dishes. In jumping around from topic to topic we would discuss all sorts of news, from the international to the family scale, picking everything to pieces. Shura loved to laugh at Mila's height (173 cm—nothing compared to today's kids!), he would move a stool up to her and climb on it, hunching down so as to remain lower than her and stretching his arms upward. He would disappear just as unexpectedly as he had appeared, so that all of our conversations finished with an ellipsis...rather than a period. Our visits to the Shal'nikovs were completely different. Shura changed completely in the presence of his wife, Ol'ga Grigor'evna, Olechka, who with her gentleness, kindness, and tactfulness was the true soul of the house at No. 2 Vorob'jevskoje Shausse and the guardian angel of the Shal'nikov hearth. In her presence Shura became gentler, quieter, more assiduous; it was Olechka who did all the work, often baking something very tasty for our visit (at her table everything seemed tasty), and Shura became himself only when washing dishes (his favorite task). At one time the hospitable Shal'nikov apartment was for us also one of the places in the institute apartment building where we could visit with Lev Landau (Dau), Eugen Lifshits, Kapitsa's private secretary Oleg Pisarzhevskiĭ, and the Petushkovs (the family of real artists of glass-blowing), and simply sit in the shade in good company on one of the benches in front of the building. But this building became empty when, in the space of only two months at the end of 1985, Lifshits and Olechka Shal'nikova died, and Shura was left alone... .

However, let me go back more than half a century, when I first heard from my parents about the existence of a remarkable relative. Our mothers were first cousins and were very close until old age, and when in 1928 my family moved from Vitebsk to Leningrad, I soon learned that here lived our remarkable relative Shura Shal'nikov, a scientist already universally recognized in spite of his youth (held up as an edifying example for a five-year-old boy). My memory of those days does not include Shura himself or his mother, Aunt Tsilya; I remember Shura's brother Grisha and his aunts, Aunt Grunya and Aunt Sonya, who lived near the Moscow gates and had the "Golden Library" of children's books, famous even before the revolution, with which I passed the hours while my mother and the mistresses of the house leisurely discussed their affairs and problems.

The member of my mother's family, the Melameds, closest to Shura in Leningrad was my mother's elder brother Solomon, a mathematics teacher. In March 1938 he was arrested and disappeared forever; apparently he was shot in the same year. After moving to Moscow, Shura became very friendly with the younger brother of my mother, Il'ya, a man with an amazing biography, very bright and interesting. During the civil war he was a commissar of the Tomin brigade in the Gaĭ corps of the Red Army and was awarded the Order of the Red Banner. He then enrolled in the Technological Institute and was sent for two years to the United States, to Ford, to acquire experience for the construction of the Stalingrad tractor factory, where for a number of years he worked as head of an assembly shop. He then became director of one of the first giants of our industry: the Sharikopodshipnik factory (the first state-owned ball-bearing factory). Shura often visited Il'ya at the Narkomtyazhprom (Ministry of Heavy Industry) house at No. 9 First Koptel'skiĭ Lane and at his dacha. He very much liked Uncle Il'ya's wife Valya, of whom he said many decades later that he never saw a more beautiful woman. From the day that his beloved Sergo Ordzhonikidze died in February, 1937, Il'ya knew that he was doomed. He was arrested nine months later, and after the Twentieth Congress of the Party we learned that he was shot in April 1938, and rehabilitated posthumously almost 20 years later.

In the summer of 1940 Shura took his vacation in the rest home for scientific workers at Peterhof, near Leningrad; his friend from Ioffe Institute,

Alexsander A. Yuzefovich, and my sister Mira, then a post-graduate student of famous physiologist Academician Leon A. Orbeli, were also there. My sister returned home completely charmed by Shura; he sent her an uncharacteristically warm letter, rejoicing in the renewed friendship (I still hope to find this letter among the hundreds left to me after Mira's death in February, 1987). Therefore, by my first student years I had already heard a lot about Shura, but I never saw him with my own eyes—I have no childhood recollection of this.

Fate brought us face to face in March, 1942 in Kazan, where I went with my mother and sister to escape the blockade in Leningrad via the famous ice road, the Road of Life, across Lake Ladoga, followed by an interminable three-week journey in a cold truck through Sverdlovsk. Shura and his family (including his mother, who had arrived in the summer of 1941 from Leningrad) lived in a hostel on Bankovskiĭ Lane. He worked in the Kazan University building, on Chernyshevskiĭ Street, on the second floor, at the far end of the corridor, where space was allocated for the Institute of Physical Problems (IPP). My mother took me there for advice on where I should go to work and whether or not this work would be compatible with my third-year studies in the chemistry department of the university. There were two possibilities: becoming a technician at a synthetic rubber factory or a laboratory assistant in the Colloidal-Electrochemistry Institute (CEIn) of the USSR Academy of Sciences, in the laboratory of Simon Z. Roginskiĭ. Without thinking, Shura answered decisively: "The factory, of course." We did not follow his advice, and for some time I was afraid to meet him, but when I did and he began to ask me about my work I realized that he was not angry.

Our laboratory was originally a chemical physics laboratory; it was transferred from the Institute of Chemical Physics to the CEIn during the evacuation, in Kazan, so all the older coworkers knew Shura well from Leningrad. My immediate boss Solomon Yu. Elovich liked to recall his great experimental skill, completely unrestrained temper, and excitability. He told of how Shura with his own hands would assemble his own very complicated glass apparatuses, and would usually finish off a seal with a long, thin curlicue. Once Shura called Yuli B. Khariton to admire a newly finished piece of apparatus, made with great effort and particular pride. And then—horrors!—in an awkward motion Khariton broke off the tip and air rushed into the apparatus, thoroughly damaging it. Shura rose up in grief and fury, grabbed a file, and ran out of the room to do a dreadful deed—he literally smashed Khariton's apparatus. A common sight was Nikolai N. Semenov shooting like a bullet out of Shal'nikov's room followed by the enraged Shal'nikov shouting "Stay away from me! And don't bring Natal'ya Nikolaevna!" However, such scenes did not interfere with the very warm and kindly mutual feelings between Shal'nikov and both the Semenovs; his relationship with Natal'ya Nikolaevna in the 1970s and 80s when both she and Shura were over 70 was particularly touching.

However, let me return to my reminiscences in chronological order. In the summer of 1943 Shura and I found ourselves in Moscow, and on August 12 the 17th birthday of Khariton's daughter Tata (Tatiana) was celebrated in the apartment house of the Institute of Physical Problems, where the Kharitons were living temporarily. Shura invited me to spend the night in his empty apartment (the Shal'nikov family was still in Kazan). We talked about our common relatives and about work, and he, to my joy and pride, suddenly said to me: "You know, I just can't address you informally—by 'ty' (you) while you address me formally—by 'Vy' (thou); let's both call each other by 'ty'." (In German "ty" corresponds to "du," "Vy" to "Sie"; sometimes people drink together some liquor for "Brüderschaft" to move from "Sie" to "du.")

In 1944 I became a post-graduate student at the Institute of Chemical Physics (ICP) and began living two steps away from the Shal'nikovs (they had all returned to Moscow in the spring of that year), first just spending nights in the accounting office of the ICP or in the laboratory, then beginning in 1946 lodging in the apartment of one of my supervisors—Nikolai M. Chirkov, in the ICP courtyard. At this time Shura obtained, in exchange for a motorcycle he had given up in 1941, a BMW motorcycle with a sidecar, captured in the war, which he named "Sakhára" and he often took me for a ride. Still now one of my most pleasant memories of those youthful years is the warm spring wind blowing in my face as Shura raced the motorcycle around on Vorobjevskoje Shausse while I rode in the sidecar. Sometimes my uncle Il'ya's wife Valya, still very beautiful at the age of 35, came to visit me from the other side of town, and Shura happily joined us and we took turns riding in the sidecar of his BMW. Right after the end of the war Shura went to Germany with a number of other prominent scientists to select equipment; he wore a kind of colonel's uniform (they were called humorous "trade union colonels"), and sometimes jokingly, sometimes seriously would say that when greeting someone of lower rank who saluted him, he politely took off his service cap.

In April 1946 Semenov celebrated his 50th birthday, and Yakov B. Zel'dovich and I wrote a humorous play "Jubilee" for his jubilee. The play was presented at the party held at the IPP. On the playbill, after the characters and actors (for example, Semenov was played by future Academician Vladislav V. Voevodskiĭ and Natal'ya Nikolaevna Semenova was played by her daughter), it was said "Staging—Aleksander I. Shal'nikov." Shura was responsible for all the light and sound effects.

In August, 1946 Pyotr L. Kapitsa fell into disfavor, and elections to the Academy of Sciences were held at the end of that year. Landau became Academician right away, and Khariton, Zel'dovich, and Shal'nikov became Corresponding Members. When I arrived to congratulate Dau, he began to analyze the results of the elections. "The most surprising thing," he said, "was that bossless Kike Shal'nikov was elected!" What Dau meant was that at the time of the elections Shal'nikov had no "godfather" Academician who could push him through; he was a railroad car without an engine. A celebratory supper for the elections was arranged in the rooms of the theory division of the ICP. At this supper I read a long poem in honor of our two new Corresponding Members; a considerable part of this poem was devoted also to the guests. In particular, it contained the lines: "But we are not willing to drink only here,/ we are awaiting the honors,/ and already preparing for/ the feasts which Shal'nikov will hold for us!" However, there was no such feast; Shura endured any sort of banquet only with great difficulty, and to play the hero at one was completely beyond him. Already to the Corresponding-Member period of Shura's biography belongs the apocryphal story of how some visitor at the IPP, seeing a puny boy in a jacket, said to him: "Boy, run and get Corresponding Member Shal'nikov!" and was flustered when the boy answered: "I am he."

Spring 1947. Shura suddenly told me that we needed to have a serious talk. He came to visit me in the Chirkov apartment and took me out for a walk. It turned out that he was burdened with a commission from Natal'ya Nikolaevna Semenova, who was worried that the romance between me and my future wife Mila was becoming serious. Shura began persuading me that we should split up, because I could not give Mila the kind of life she was used to. He spoke somewhat unwillingly, and his words seemed not to be his own. I proudly assured him that I could provide for Mila, after all, I was already a Candidate of Science! Shura did not persist, and our conversation quickly moved to other topics.

During those years I had a special interest in the Shal'nikovs also because Shura become one of my first acquaintances with a television and a car. He got the television—a homemade one, with a tiny screen—in 1947, and the foreign car somehow appeared even earlier. Shura liked very much to play with his two new toys, and with great satisfaction showed them to everyone interested. In addition, when he bought various books at the Academy of Sciences bookstore, he tried every way possible to quickly get rid of those he had bought rashly, and often loaded me down with various junk books before leaving. I remember one such present from him which long occupied space in our bookcase: a very fat book by mediocre writer Aleksander Popovskiĭ entitled *The Rules of Life* with panegyrics to Trofim Lysenko, in which Shura wrote to me the dedication: "That's what you have to have."

After Mila and I got married in 1947 and our eldest son, Mitya, was born the following year, my mother and sister Mira began to visit us every summer. I remember our visits, with Mama and Mira, to the Shal'nikovs, where

Mama would go up to Aunt Tsilya's room, and Mira and I with Shura and Olechka would go over the fates of common relatives and acquaintances, and common reminiscences about the topic of the day. Many conversations were about the notorious "Pavlov session" of 1950, which played a tragic role in the fate of Mira's teacher Academician Orbeli and made Mira lose her job. Shura would very indignantly, with anger and sarcasm, review the black pages in the life of our science and our country. However, March 1953 was approaching... .

In the middle 1950s, during the time I was at the Lebedev Institute (1952–1961), when we were preparing synchrotron experiments on the Compton effect on the proton and on neutral pion photoproduction, I turned to Shura for help in working with liquid hydrogen, and his help was truly invaluable. Almost every day I went to the new cryogenics building in the courtyard of Moscow University, where Shura's low-temperature group was located. He taught me how to glue hydrogen targets from styrofoam, how to inundate them with liquid hydrogen from a Dewar vessel, how to monitor the level of hydrogen in the target placed in a bremsstrahlung radiation beam, and many other important things. I then was proud when all the first experiments with liquid hydrogen at the 265-MeV synchrotron at the Lebedev Institute could be carried out only with my participation; for some time no one but me had all the necessary skills, and I taught them to our entire group personally. Therefore, Shal'nikov became the grandfather of all the Lebedev Institute experiments involving liquid hydrogen (for which we thanked him in our papers published in *Doklady Akademii Nauk*). Soon Abram B. Fradkov arrived from the Institute of Physical Problems and a real hydrogen station was built.

In May, 1955 the 50th birthday of Shura was celebrated very merrily in the cryogenics building at MSU, and among other selections we played "Lectures on liquid air," which I had taped (that year we all acquired "Dnieper" tape recorders and copied from each other the wonderful records of Oleg Pisarzhevskiĭ). The tape of this lecture still exists, and the text of the lecture was reprinted almost without cuts on page 16 of the *Literaturnaya Gazeta* in 1975 for the 250th anniversary of the USSR Academy of Sciences, when the gazette acquainted its readers with the humor of scientific workers. A wonderful story by Shal'nikov about a typical seminar at the Institute of Physical Problems was also published there.

It must be said that stories told out loud by Shura with his characteristic, rather muffled patter were very witty and evocative, juicy and full of hyperbole and humor. One of Shura's favorites was the story of how Semenov and Igor' V. Kurchatov "killed a little old lady." This is how Natal'ya Nikolaevna describes this event.

"Back in 1927 a group of young scientists, at that time quite unknown but connected through their work at the Physico-Technical Institute and very good friends, decided to spend summer vacation together. Some acquaintance advised them to go to Lake Seliger and look up the priest of the church Nikola Rozhok. They did this. The place turned out to be really wonderful, beautiful as in a fairytale, and also nearly uninhabited: on the steep, precipitous bank directly over the lake stood an old church and next to it three houses, belonging to the priest, the deacon, and the psalm-reader. The nearest village was several kilometers away. Fishing and hunting were guaranteed. The Leningraders who had arrived, the Semenovs, the Kurchatovs, the Strelkovs, and Anton K. Val'ter, settled in these three houses. The vacation was wonderful: swimming, boating, sunbathing on the beautiful white sand, and, especially, being with friends. But something happened. On the day of Prophet Il'ya a service was held in the church, with people from the nearest village attending. An unusually violent thunderstorm unexpectedly arrived. We all ran to hide in our houses. But Pyotr Georgievich Strelkov and his wife Vera Anatol'evna continued to carelessly sit on the sill of their open window. Their three-year old child was between them. An old woman who had been on her way to the church was leaning against the outside wall of their house; she apparently sought cover from the rain. A loud thunderclap was heard. Lightning struck the Strelkov house. It burned part of the roof and the cloth on the table near the window. Vera Anatol'evna fell down unconscious, and on the skin of Strelkov, who was bare to the waist, appeared a sort of branched mark from the electric discharge. The child was sitting on the floor unharmed. Only the poor old woman lay without a sign of life by the wall of the house. Semenov and Kurchatov ran up in response to cries. Someone began to give advice: a person struck by lightning should be buried in earth. The two young scientists began to energetically bury the old woman. But alas! All their efforts were in vain. And even worse: the old woman seemed to show some signs of life (in any case her eyelids started twitching), but after these measures it became clear that she had departed for a better world... . Strelkov's wife, who could not speak for several hours, successfully recovered. Strelkov himself for several days proudly showed everyone his tattoo made by the lightning. For a long time afterward the scientists who, as it were, were the undoing of the unhappy old woman, liked to remember this event and banter about it. Shal'nikov especially loved to talk about this when he met friends."

One of Shura's favorite expressions in stories about his various debates and quarrels was "I am 200% correct!" When our second son Andryushka was born, Shura liked to hold him up naked and say "How brand-new he is!" A wonderful way of describing a baby!

It was Shal'nikov who gave Kapitsa the nickname "centaur" by which he has long been known around the world.

Typical of Shura were categorical judgments about many common acquaintances (the denigrating epithet "careerist!" often came from his lips). It must be admitted that in most cases they in the end turned out to be true, even when they at first seemed to me too strong or completely unjustified. Shura was like gunpowder, sometimes he set himself on fire with his own monologues. Olechka would then try to calm him down, gently opposing him; they complemented each other perfectly.

On January 7, 1962 Semenov and I were sitting at the table in his apartment when the telephone rang. It was Shal'nikov, very upset, informing us about the accident that Landau had been in, and asking for thiourea to be procured urgently. Semenov undertook to phone up chemists, but already the first night Robert Maxwell sent the needed preparation by plane from London. The dramatic, epic struggle for the life of Dau had begun.

In the Shal'nikov house there was a place for preparing food for the patient. Dau needed to follow strict dietary rules, and the food had to be in liquid or semiliquid form, because he had to be fed by an enema, as he was unconscious. Olechka spent whole days at this very important work and Shura helped her; dozens of physicists several times a day took food to the hospital in bottles and brought the empty ones back to the Shal'nikovs. This went on for several months, until Dau could eat by himself.

In the fall, when I returned to Moscow, Shura took me with him to the Burdenko Institute, where Dau was staying. He was lying silently on the bed, and seemed to me to be completely indifferent to everything. There was another man in the room, whose face looked familiar to me, but I did not remember his surname. "Who is that?" I asked Shura, after he had left. From the bed came the quiet but distinct voice of Dau: "A very weak physicist." Shura and I were delighted, thinking that the old Dau would be back with us; given enough time, he would get strong. Alas! As is well known, these hopes were not realized.

Of course, I do not at all pretend that these reminiscences are complete. It is difficult to reconstruct the images of the past in a definite (chronological or some other) order; they come unexpectedly, at random, usually because of some association. For example, I suddenly remember how afraid Shura was of moths or was squeamish about touching them. And how he led me on with stories about Mila's wild driving and demanded that I not allow her to take the wheel... .

Finally, let me tell the complete story of Shura's election as Academician in March 1979. When the start of the election campaign was announced he was in Canada, visiting his youngest daughter Tanya. His oldest daughter, Natasha, whom Kapitsa had contacted about the nomination at the Institute of Physical Problems of Shal'nikov as candidate for Academician, phoned me and asked how the necessary papers could be collected; if papa were here he certainly would not begin to do it. Nevertheless, I told her that in my opinion the papers should be gotten ready, and she quickly began helping to collect them. Shura returned to Moscow only a short time before election day. As is well known, Anatoly P. Aleksandrov gave a very warm speech about Shal'nikov to the assembly of the Division of General Physics and Astronomy, about his important role as teacher of many generations of our physicists in the art of experiment, and, if I am not mistaken, Shura was elected in the first round, completely unexpectedly for himself and others. When in the evening I phoned him to find out who had been elected in their division, he answered in a very embarrassed voice, "Me." Between the Division election and the General Assembly election on the following Sunday, Mila and I took Shura off to Krylatskoe, where our son Mitya was demonstrating his hang-glider flying. The very first flight ended with him landing on his back in a tree and lying in its branches helplessly, like a beetle, waving his arms and legs, but unable to turn over. Shura immediately came to his rescue, and in doing so got so scratched that his face was bloody. Wiping him off with a handkerchief, I said that I would keep it as a souvenir of the first blood of Academician Shal'nikov.

We mentioned official papers of a different type. Shura was childishly helpless at writing them. One of the last clashes related to this that I remember was a telephone call from him after the meeting of the Committee on the Lenin and State Prizes of the USSR, where at the suggestion of Aleksandrov Shal'nikov was added to the list of candidates for the State Prize, nominated for the application of low-temperature techniques to surgery. Shura was really alarmed: he thought that this list was almost entirely made up of people who had not done anything and that none of the most deserving people were on it. He therefore did not at all want to win a prize in such company and was very worried about withdrawing his candidacy before it was too late, and asked where and how this could be done. As far as I remember, Shura's agitation produced an effect: the prize was awarded only the following year, to a different group of authors, in which he could regard himself as figuring in.

Shura was laid low by his unhappy fall from a ladder in October 1983, when he remained alive literally by a miracle, but suffered greatly. Mila and I were not in Moscow at the time; upon our return in November we found him completely unrecognizable, reacting slowly to everything around him. He came to himself and returned to his former aspect very gradually, but nevertheless we were happy to recognize in him more and more of the characteristic features to which we had become accustomed over many decades. He began spending most of his time with Olechka at the dacha in Novo-Dar'ino, and they were both very happy leading a quiet and comfortable life there.

Shura's 80th birthday in May 1985 was a happy one. But that year ended tragically for the Shal'nikovs: exactly on New Year's Eve Ol'ga Grigor'evna died very suddenly. Her death overpowered Aleksander Iosifovich. Many times he asked me, "Vit'ka, where is Olya?" And I simply did not know how to answer; it was as though he really couldn't understand what had happened.

However, it would not be good to end these reminiscences on a melancholy note. In my memory Shura remains ageless (as he felt himself to be, at least until he was 75), full of energy, enthusiasm, various scientific plans, in perpetual motion, like a drop of quicksilver, continually arguing, strong in his opinions, honest, extremely accurate in his sympathies and antipathies, unique and irreplacable.

Some Pages from the Past

(Vladimir Iosifovich Veksler)

As the years go by, one realizes more and more how shockingly brief was the life of Vladimir Iosifovich Veksler and how young he remained essentially up to his last days.

He was only 44 when I first saw him in the winter of 1951 at the "Pitomnik," (the name inherited is from a plant nursery located here before the construction of the accelerator) where I had gone to ask for permission to use the 265 MeV synchrotron to measure the photoneutron yields from various thick targets.

A bit short of our destination the car got stuck in the mud, so it was necessary to travel the last few dozen meters on foot. I therefore entered the office barefoot and was quite embarrassed. But V.I. laughed heartily at the incident and immediately led me to a sink to wash my shoes and socks. Then the business discussion went very easily. V.I. quickly grasped the essentials of the problem, asked about the results of our experiments at the Dubna synchrocyclotron, sent for several of his colleagues, to take care of our group and clearly was immediately filled with the spirit of competition and the desire to show that their accelerator was better than any other also for our particular problem. Our experiments went better than we had hoped and were completed within a very short time, largely thanks to V.I.

After a year it happened that I temporarily left the Institute of Chemical Physics and went to work for nine years in the laboratory of V.I. Veksler at the Lebedev Institute; at that time it had the disorienting name of Standard laboratory.

The first five years of my life at the Lebedev Institute were filled with frequent meetings with V.I. and discussions with him about various scientific and practical problems; then V.I. concentrated nearly completely on his work at Dubna.

I was most struck and, above all, delighted by V.I.'s extraordinary capacity for hard work combined with rare talent and obsession with research. He possessed the quality, rather rare nowadays, of insatiable curiosity about science, a compulsion to examine the experimental apparatus with his own eyes and hands, to personally hunt down the theoretical work and check that it was sound. The presentation of reports at seminars at which he was present and discussions about work in progress were very interesting and at the same time difficult, and sometimes involved heated arguments.

Here also was manifested the quality of V.I. ("one shouldn't leave the words out of the song") which I found most difficult to get used to. He was not impartial; he was lavish with praise for ideas or work which came out of his institute, his laboratory, his school, without worrying much about the adequate evaluation of the quality and value of other research. In all discussions there existed a sort of presumption of correctness on his own side and a presumption of error on the other side. Having come from a different institute, a different school, I strongly felt this. As a result, the interest and attention that V.I. showed toward the work of our group became more precious. Spring and summer of 1956 came, memorable for the first international meetings after a long hiatus. We busied ourselves for whole days with styrofoam targets for liquid hydrogen, preparing experiments on the Compton effect on the proton and on π^0-meson photoproduction. These were the first studies carried out using liquid hydrogen at the Pitomnik; V.I. was very glad that they were finally being carried out and wanted, as usual, to verify this with his own eyes and hands. He arrived at a very unusual time and became touched when he saw the entire group sitting and watching the lengthy filling of the target with hydrogen and the placement of the target in the beam. And when, at the High Energy Physics Conference in Moscow, Weisskopf arrived and described to V.I. how David Frisch was also working on the Compton effect on the proton, V.I. arranged for the three of us to meet and asked in detail what was being done in Cambridge; here I clearly felt that V.I. was our fan. In August our entire group went away for vacation exhausted but satisfied; before leaving I showed my colleagues a note from V.I. to Igor' Evgen'evich Tamm, in which he asked him to submit our two papers to *Doklady Akademii Nauk* and vouched for their high quality.

The 50th birthday celebration of V.I. in March 1957 became to some extent his farewell party upon leaving the Lebedev Institute—he was nearly completely focused on his work at Dubna. Our life in the Pitomnik became quieter, but much less interesting.

Another six years passed, and V.I. was elected as the head of the new Nuclear Physics Division of the USSR Academy of Sciences. Now for V.I. all of our nuclear physics became "his own," and he was ready, like an ardent missionary, to spread the nuclear religion among pagans like chemists and biologists. By that time it was already several years since I had returned to the Institute of Chemical Physics (ICP), where I had set up a laboratory of nuclear and radiation chemistry. Now I became a sort of nuclear ambassador among chemists, and in the spring of 1965 V.I. asked me to help him organize a two-day scientific meeting between two divisions, nuclear physics and general and technical chemistry, devoted to applications of nuclear methods in chemistry. Right after this he wanted to leave for Chernogolovka (scientific town at approximately 60 km from Moscow): it was necessary to choose a site for the future meson factory, and by that time many buildings of the ICP had already been constructed there.

The three-hour trip by car there and back went unnoticed. V.I. recalled the old Lebedev Institute, trips to the Pamir, and made plans for the construction of new large high-energy physics laboratories. He inspected the two building sites and after dinner visited several ICP laboratories. It was already evening when we returned to Moscow. V.I. was satisfied with the trip, but decided that it would be better to build the meson factory closer to Moscow, toward the southwest. As is well known, Pakhra was chosen.

V.I. considered Mössbauer spectroscopy a very promising line of research. Now in our country there are many institutes devoted to this field, and they should be aware of the very important role played by V.I. at that time in fostering this development. He commissioned me to prepare a report on Mössbauer spectroscopy and arranged with Mstislav V. Keldysh to present it to the Presidium of the USSR Academy of Sciences. Together with Mikhail D. Millionshchikov he prepared a plan which incorporated a number of important and urgent measures and intended to take part in the discussion himself. The Presidium session went well for a change and ended not only with Keldysh supporting all of V.I.'s proposals, but with charging me with the task to present the report a second time to a larger audience, at the annual General Assembly of the USSR Academy of Sciences on February 8, 1966.

Alas, V.I. was not present at either the Presidium session or at the General Assembly of the Academy: at that time he was hospitalized in Leningrad after suffering an infarction. In letters to him I described in detail how things were going, warmly thanked him for his help and for putting Mössbauer spectroscopy "in orbit," and wished him a rapid recovery.

However, I was not destined to see him again.

Forty-five Years
(Yakov Borisovich Zel'dovich)

I am beginning to write these reminiscences of Zel'dovich on March 8, 1989, which would have been his 75th birthday. I just saw and heard him for the first time since his death: the film "The scientific personalities of Academician Zel'dovich" was shown on television, and again his death seemed completely outrageous and unjust.

Forty-five years passed from the time I first saw Yakov Borisovich in Kazan to the evenings of November 29 and 30, 1987, just before the last evening of his life, when, with his usual youthful lightness, he ran up the steep stairs to visit us, to look at some recent American journals with articles about his studies from the 1930s and 1980s and to discuss the approaching elections to the Academy. In thinking over this long but brief period, which coincides with my entire independent life, in sorting through Yakov Borisovich's notes, his humorous verses, and various notebooks, I feel particularly strongly the emptiness left by his unexpected death.

I first heard of Yakov Borisovich (YaB) from one of his closest friends and colleagues, Ovseĭ Il'ich Leĭpunskiĭ, right after I arrived in Kazan from besieged Leningrad and began to work as an assistant in the laboratory of Simon Z. Roginskiĭ in April 1942. There is a lot to tell about life at the Institute of Chemical Physics (ICP) during the evacuation (the Forties, fatal years, years of lead, years of gunpowder; the war going around in Russia, and we so young ... —from my favorite verses of David Samoilov), but it will have to wait for another time.

Leĭpunskiĭ spoke of Zel'dovich with undisguised delight, with the highest praise, and it was difficult to connect the figure from these stories with the fast-moving, bespectacled young man I met on the stairs, in the library, in line at the bread pantry, whom everyone addressed by his first name, without ceremony. In November I heard YaB for the first time at a session of the Scientific Council of the Institute of Chemical Physics; this was the first dissertation defense I had witnessed. The 34-year-old Aram B. Nalbandyan defended his doctoral dissertation, and the 27-year-old Nikolai M. Émanuél' defended his candidate's dissertation. However, all the opponents—there were three, as for doctoral defenses—were in favor of awarding the doctoral degree to Émanuél'. His work, which demonstrated the decisive role played by the intermediate product sulphur monoxide SO in the oxidation kinetics of hydrogen sulfide, was certainly distinguished, and Nikolai N. Semenov very insistently supported it as doctoral-level work. I think that it was just his persistence that generated opposition from members of the council, and YaB became one of the two leaders of this opposition. Several times he took the floor and objected to the fact that the defender of the dissertation had skipped the candidate level, citing as arguments the names of leading chemical physicists who had not yet become doctors. My inclination (and sympathies) were clearly on the side of Émanuél', and after the council had pronounced him a Candidate of Science and Semenov from the empty corridor of the Institute phoned Pyotr L. Kapitsa and bitterly complained about the result of the vote (I was on duty at the Institute that night and involuntarily overheard him), I fully sympathized with him.

To wrap up this story—and to edify today's young scientists—I add that later my two friends and neighbors at home, Zel'dovich and Émanuél', had a close and warm relationship for many decades, which ended only with Émanuél's death in December 1984.

1943 arrived. YaB, like Semenov, Yuli B. Khariton, and other heads of the Institute of Chemical Physics, visited Moscow more and more often. The atomic problem arose, and there was discussion of moving the entire institute to the capital. Much laughter was produced at the Institute of Chemical Physics in Kazan by the jokingly angry telegram of YaB, urgently requesting needed materials: "Nonreceipt of reports upsets your mother. Rearrange. Zel'dovich." (Rearrangement attaches here some indecent sense.) YaB had a great sense of humor, and the reader of these reminiscences will have more opportunity to see this.

In the spring of 1943 YaB was awarded the Stalin prize for his work on the theory of combustion and detonation. The awarding of such a high prize to so young a person (29 years old!) and to an individual rather than a group was exceptional at that time.

In the summer of the same year at the call of my boss Roginskiĭ I also was sent to Moscow, where my formal acquaintance with YaB soon began. Most of our colleagues from the Academy of Sciences were still in evacuation; only a few were in Moscow, and we all ate together using ration coupons at either of the two academic cafeterias, so we got to know each other well.

In the fall the entire family of Semenov arrived in Moscow. The Semenovs lived for over a year in the Moscow Hotel (room 1211) and from time to time I met YaB when I went to visit Semenov's, my friend Yuri.

In February 1944 we all attended a concert by Aleksander Vertinskiĭ, a sensation at that time, at the Scientist's House. I learned to imitate his singing tolerably well, and YaB fairly frequently asked me to perform songs that he liked.

Sometime in the summer of the same year, having accidentally met in Academy's health resort near Moscow—Uzkoe while visiting Semenov, we spent an hour and a half returning together to the institute, by tram and on foot. Our long conversation about Leningrad, the war, books, and common acquaintances brought the first hints of warmth into our relationship. At the end of 1944, when I was already a post-graduate student of Semenov, I settled in the institute building then under construction, first sleeping in the accounting office and then in the laboratory, where at night a folding bed was placed next to the laboratory bench, and the day began with turning on the electric stove (for morning tea) and the vacuum pump. YaB frequently came to visit me, and our conversations at that time were roughly equally divided between science and the fair sex. As a first test in science he set me the problem of finding and correcting all the errors or misprints in his article with Yuri A. Zysin on convective heat transfer in an exothermic reaction in a jet. As far as the topic of women was concerned, many years afterwards Vitaly L. Ginzburg teased me about how he had first heard my name. He had been walking with YaB in the corridor of the only wing of the ICP in use at that time, and some young female colleague had asked YaB in which room Gol'danskiĭ was working. "That one," said YaB pointing out the door, "but bear in mind that a young woman entering that door can be considered to have gotten married" (I cannot help recalling here some lines of a poem I like by Yunna Morits: "When we were young and talked beautiful nonsense ...").

The unforgettable days of May 1945 passed, and in June the USSR Academy of Sciences celebrated its 220th anniversary. The Joliot-Curies, Langmuir, Hinshelwood, Svedberg, Auger and many other eminent guests came to our institute. YaB (who, incidentally, received his first decoration at this time, the Order of the Red Banner of Labor) was one of the main attractions for the guests. At the entrance to his laboratory two of the guests stopped at the door, each to let the other go in first, and the whole procession came to a halt. YaB had spoken German well from childhood, but it was not customary or even desirable to speak in that language at that time. In broken English he laughingly reminded his guests of the famous dialog between Chichikov and

Manilov in Gogol's "Dead souls" in a similar situation. Two of the guests, without listening to the end, pushed ahead and, of course, became wedged in the doorway; YaB triumphantly ended his commentary with the words: "It happened just like that in Gogol."

The Zel'dovich family, which after two daughters finally got its long-awaited successor Boris, now a Corresponding Member of the USSR Academy of Sciences (YaB used to joke "one, two, three, go" about this), settled on the first floor of the apartment house in the courtyard of the ICP, in apartment 11. Fairly often YaB invited me over for a bite to eat and would say to his little daughters, Olya and Marina, who were watching this procedure with curiosity, "Do you think he really is eating? Silly girls, he is just faking." This expression (for example, "let's fake drinking tea") was a common joke between YaB and me for a long time.

At that time I was worrying about writing my first paper, "Indication of chain reactions by nitric oxide" (which appeared in *Uspekhi Khimii* in 1946) and couldn't figure out how to begin it. When he learned of this, YaB said, "Nothing could be simpler. You must begin with the words 'The vigorous development.' What will the article be about? Oh! Then you must begin 'The vigorous development of methods of indicating chain reactions ...'."

In late 1945 or early 1946 YaB turned to me with an unexpected request: to jointly write a comic play to be performed at the 50th birthday celebration of Semenov coming up in April 1946. We enthusiastically set to work. We worked together on some scenes; we wrote others separately and then judged each other's work. YaB wanted the play to have as strong a "sexual-Freudian" tone as possible. The play depicted (comically) the events of 1924 (Semenov's marriage) and 1926 (discovery of a lower limit on the ignition of phosphorous by Khariton and Zinaida F. Val'ta) and the years up to 1986 (the final scene was entitled "Forty Years Later"). One of the several interludes between the two main acts of our play was devoted to the first appearance of the young Yasha Zel'dovich at the ICP. Here I give this interlude with little explanation; the character "Unknown" appeared with the request to "call the glassblowers to order" in all the acts over the course of several decades.

Interlude I. 1931.

SEMENOV (Sitting in his office). (A boy in glasses enters).
BOY. Uncle, are you the director?
SEMENOV. Yes, What do you want?
BOY. I want to become a professor.
SEMENOV. Oh really? And what is two times two?
BOY. Why are you laughing, Uncle? Give me an integral to do! Look: you haven't written the subscripts right!
SEMENOV. Yes, it's true, they are wrong. Interesting. What is your name, boy?

BOY. Yasha Zel'dovich.
SEMENOV. Well, come with me to the scientific council! (Goes to door. Meets Unknown.)
UNKNOWN. Nikolaĭ Nikolaevich! Call the glassblowers to order!
SEMENOV. Later, later! Let's go now to the scientific council!
UNKNOWN. There will be no council. I couldn't get anything on equilibria
SEMENOV. Disgraceful! You haven't learned yet how to compute equilibria!
BOY. So grown up and he still can't compute equilibria.
UNKNOWN. (Gives the boy a slap. Exits.)
(Boy again goes up to Semenov.)
BOY. So will you take me to the professors?
SEMENOV. OK, we'll talk about it later.
BOY. (Crying). No, take me to the professors.
SEMENOV. Actually, it is necessary to advance young people. OK, I'll take you to the institute.
BOY. I'll come tomorrow. I'll bring Doda Frank (i.e. David A. Frank-Kamenetskii). He's also good. We'll play leapfrog together. (Runs off. Shouts loudly: "I'm going to be a professor! I'm going to go tell the guys!")

Having taken on the responsibilities of playwrights, YaB and I also became the directors, gathered together a company, and conducted some rehearsals. The role of Semenov was successfully played by future Academician Vladislav V. Voevodskiĭ, his wife Natal'ya Nikolaevna was played by the Semenovs' daughter Mila, and Yasha Zel'dovich was played by the son of one of the oldest colleagues at the Institute Bronislav S. Kogarko, who is now a Doctor of Science. We did not decline to become actors. YaB appeared in the scene "Morning of a post-graduate student," in a dialog between the post-graduate student (clearly understood to be the present author) and Semenov. This scene was almost completely written by YaB himself, and fully conformed to his desire for strong sexual-Freudian overtones. Our play was successfully presented on April 21, 1946 in the auditorium of the Institute of Physical Problems, and the lighting and sound effects were provided by Aleksander I. Shal'nikov.

I devoted the last lines of the concluding scene of the play to a report of the life and work of Semenov, read forty years later on his 90th birthday in the presence of YaB. Alas, Semenov died in that year and Zel'dovich in the following one, and now the lively couplets arouse in me only sadness for the irreversible passing of time and people.

In the fall of 1946 I turned in my typed candidate's dissertation, and Semenov on the spot decided to transfer me from catalysis to a nuclear topic and to have me temporarily work half-time in the theory division, under the direction of YaB. I remember election day in the Academy at the end of November, when YaB was a candidate for Corresponding Member. He nervously kept going in and out of the room, disappeared for an hour and a half, and again returned. Every time I asked him, "Well, how is it going?" ("Nu kak?" in Russian) until he angrily answered, "Go and dirty your pants" ("Kakajte pod sebya" in Russian). The elections went successfully, and a few days later on the third floor of the institute, where the theory division would soon be located, a banquet was held. Here I quote a few lines from the humorous poem I read at this banquet:

> What a triumphant moment!
> Zel'dovich is a Corresponding Member!
> And Khariton has become one too.
> Let a chorus sing their praises.
> ....
> Having split up our problem,
> Now let us take up Yasha.
> He is witty and beautiful.
> He is pleasingly plump,
> And always dangerous to women,
> Even though by his weight rather than beauty.
> After giving up his pacifier, at two years
> This amazing boy solved integrals.
> He was a Candidate at twenty-two
> (Now the devil himself will not bother with him),
> He became a Corresponding Member at thirty-three.
> But nevertheless, Yasha, you should not
> Stick your nose in the air;
> After all, Christ beat you.
> He was just the same age as you
> When he ascended to heaven.
> (Here to make a rhyme with this line
> We remind you of Golgotha!)
> No, you will not catch up with Christ,
> Though there are nails there is no cross.
> There is no need to ascend to heaven,
> There are things to strive for on earth.
> The Institute's alcohol into the glasses!
> Drop that flirtation with the outside!
> Our toast to our pride,
> Our toast to Yasha!
> Our Yakov can be anything:
> Scientist and volley-ball player,
> Playwright and tennis player,
> In Kazan an anti-aircraft fighter,
> In short, our fine fellow Yasha.

In 1947 I spent several months in the theory division of the ICP, where many brilliant people were working, each on his own topic, under the direction of Yakov Borisovich: people like Frank-Kamenetskiĭ and Aleksander S.

Kompaneets, quite young and talented people like Seryozha D'yakov (who died tragically three years later) and Kolya Dmitriev, and also Erna A. Blyumberg, who helped with computations.

The division published a "dvergazet" (a door-newspaper; almost a literal translation of the Chinese word "Datsybao" which later became sadly familiar to us). Each person composed something, in verse or in prose, and posted it on the door to be reviewed and discussed by all.

I managed to win the Mikhalkov (author of Soviet Hymn) laurel of local importance by composing the following Hymn of the Theory Division:

> Our division is the hub of the institute,
> Longed for by other people,
> Our great Corresponding Member
> Defecates hundreds of ideas a minute.
> We bring the light of science to the plebeians,
> Who are always working so hard on their experiments,
> We are not afraid of Semenov,
> There are none more daring than our boys.
> Landau himself makes the sign of the cross over us,
> We are mighty, healthy, strong,
> Only our asses suffer
> When we prepare classified reports like pancakes.
> Glory to the physicists, glory to the poets
> Surrounding Yakov's throne.
> We feel good in the peacefulness of the office,
> Our balls are not scared by neutrons.

All sorts of practical jokes, sometimes not completely innocent, went on. Somehow reprints from a foreign journal, from an author unknown to us with a Russian surname, arrived at the institute, addressed to YaB. Kompaneets and I wrote on them "To my dearest friend YaB from Headquarters Captain (Rank in Tsar's army) Surname" and put them on YaB's desk. It must be realized how dangerous it was at that time (especially for a person doing top-secret work) to have any contacts with emigres, particularly former officers of the white army, in order to understand how upset YaB was and how sorry we were afterward for this joke.

Our entire life was concentrated in the small world of the institute: our living quarters were in its courtyard; we played volleyball on the grounds, YaB, Frank-Kamenetskiĭ, Voevodskiĭ, and I went for morning runs, and we played catch.

YaB participated in sports, as in science, with great passion, sometimes pushing himself to a state of exhaustion.

He loved—not without my influence—ball dances, and we quite often (sometimes every Saturday) went for a whirl at the ICP or the Institute of Physical Problems nearby. YaB obviously took pleasure in dancing, but, in spite of my persuasion, did not try to learn how to dance correctly; he simply jumped around to the music, deliberately avoiding any of the standard steps. Dancing was for him like a game of catch, and with pleasant partners what can be better!

In October, 1947 YaB came to my aid in one of life's difficult moments. Mila Semenova and I had to tell her parents that we had gotten married, something which they (especially my future mother-in-law) were not at all in favor of. YaB solved this problem. He took Semenov to his office, talked to him for a long time alone, and finally came looking for me, satisfied and smiling, so that suddenly a great weight was lifted from my heart. And, in fact, my next (very) serious conversation with Nikolaĭ Nikolaevich turned out to be as warm as always before.

YaB left Moscow more and more often to go to "X" (now we know it as Arzamas-16), and it gradually become more correct to say "YaB comes to Moscow less and less often." I do not say if January 1948 was the end of the Moscow period of his life or the beginning of his "X" period, but this month reminds me of the death of famous Jewish actor and civic leader Solomon Mikhoels. YaB came to my room without his usual smile, looking stern and worried, and suggested we go to the Jewish Theater, where the farewell to the deceased, who we soon learned was in fact murdered, was being held. Getting off the tram near the Nikitskiĭ gates, we saw a huge crowd jamming the streets in front of the theater and were taken aback. YaB gloomily joked: "Getting in without standing in line is here impossible—they're all Jews," and searched in his pocket for the card stating he was a Stalin Prize laureate. This card paved the way for us, and not only into the theater, but all the way up to the honor guard, and we could not help noticing the huge bruise on the temple of the deceased. We went back in silence, oppressed with sad thoughts. Jumping ahead, I say that the problem of the steady rise of anti-Semitism in the country, clearly encouraged from above, disturbed YaB very much and made him very bitter. Anti-Semitic attitudes were clearly perceptible behind the scenes in the attacks on YaB's book *Higher Mathematics for Beginners*. Two days before his death he told me about a nasty anonymous letter in which, in addition to the usual anti-Semitic abuse, he was called Satan. Feeling that during the years of stagnation there were serious people in high circles who were lovingly cultivating future neo-Nazis, YaB and Aleksander N. Frumkin wrote to Yuri V. Andropov on this topic, met with Sergei P. Trapeznikov, and asked for the admission to Mikhail A. Suslov and Mikhail V. Zimyanin, but all in vain.

However, let us return to the chronological story. Beginning in the fall of 1947, the small group assigned to me was commissioned to study the passage of neutrons through the shell of explosive material which was to surround the nuclear core of the future "device." We mimicked the composition of the explosive and performed experiments for over a year, first with a neutron (Ra-Be) source, then at the reactor and cyclotron of Laboratory No. 2 (later the Kurchatov Institute of Atomic Energy). Once YaB brought a group of generals to see these experiments. In complete seriousness, without a shadow of a smile, he explained to them: "In here they put little sheets made of indium and therefore called indicators," and teasingly winked at us.

Our minds usually select the more happy and pleasant memories. Actually, times became more and more base and difficult. One after another, decrees were issued by the Central Committee of the CPSU(B) on literature and art; in science, courts of honor were held for Klyueva and Roskin, and Parin, Balandin, and Parnas were sent to prison. The decision was made to turn the courts of honor into a public show (so to say "comrades' lynching court"), and at the ICP YaB and Leĭpunskiĭ were chosen as the most suitable defendants. They were supposed to confess to having attempted to force through, as an open publication, some study of theirs on gunpowder, hastily cleared by security, and Mikhail A. Sadovskiĭ (so to speak, the less guilty of the culprits) was accused of having approved the submission of this study to the press as deputy of the director. However, YaB stood up strongly for himself and refused to admit any guilt. This shameful "court of honor" hastened YaB's official departure from the ICP and his transfer to "X" to work under Khariton. When a year (or less) later, after the first Soviet atomic bomb explosion (August 29, 1949), a golden shower of praise rained down on the nuclear scientists, and the breast of YaB was decorated with the first of his three Hero of Socialist Labor stars, he specially invited to a party at his house not only all his friends, but also his worst persecutors. I will not name them here, not because of the principle "aut bene, aut nihil" of those who have departed this life, but because of the blameless children and grandchildren of the loyal lackeys of Stalinism in our institute who are still living. When there was talk of a forthcoming apology to YaB and his recall to the ICP, he quoted to me a couplet he liked from the risqué rhymed couplets "Allaverdy": "The hussar is the most handsome of all the soldiers; The sheet cannot be pushed backwards."

The time that YaB was transferred to "X" coincided with the attempt to assign so-called secretaries to him (three KGB agents who took turns being on duty, relentlessly, 24 hours a day, watching the leading nuclear scientists). Incidentally, among these secretaries were people who became genuinely attached to their wards and eventually became their indispensible aids. However, YaB (like Lev D. Landau) categorically rejected the attentions of the KGB. For several nights some unknown people spent the night on our staircase (starting in 1949 we lived on the same landing, our doors were right next to each other), standing or climbing up on the windowsill (at first my wife and I didn't understand who they were); YaB would not let them into his apartment. In the end, the secretaries could not endure such working conditions, and YaB won.

In the 1950s fate separated us. Yakov Borisovich was in "X," and in 1950–51 I was working in Dubna on experiments on the absorption and multiplication of high-energy (in those days 120 and 380 MeV was "high") neutrons, and in early 1952 because of the fight against nepotism I was transferred from the ICP to the FIAN (Lebedev Institute), where I worked for nine years. Here I must say that it was thanks to the Lebedev Institute that I was able to expand my knowledge and interests and was later able to open new scientific contacts with YaB. However, this was several years later. Meanwhile, our meetings, even though we were next-door neighbors, were rare. Being in the midst of things, YaB clearly felt the growing severity of the regime of the later Stalinist years. He keenly suffered from the troubles brought on by the Korean War, although at the time he did not like to talk about it. Only many years later did he tell me about his agonizing difficulties: if our first atomic bomb had not been made in 1949 (with the active participation of YaB), the Korean war would not have begun in 1950. And strongly sensing the danger of being spied upon, he severely reprimanded Kompaneets and me for our love of speaking too openly. Upon one of his arrivals (on the eve of which there was a small fire in our courtyard), I responded to these reprimands with verses from which our stupid carelessness and inability to understand the full seriousness of the danger to YaB literally jump out:

It was not without reason that Moscow
Greeted Zel'dovich with fire the other day,
Yasha has been honored enough,
He has enough gold stars,
And, they say, what nice stars they are!
They have begun to sing songs in Russia
About our eagle.
Still not suffering from hemorrhoids,
He instead has become a worthy hero
And can put down all the sceptics.
For the younger generation
You are a paragon of good behavior,
We will remember your lessons,
Our dear Corresponding Member.
Henceforth we will not write any verses
Or tell jokes to people,
We will try to look serious.
You, eradicator of thoughtless chatter,
Are the marvel of Rus' and all of Europe.
He has a cool head and persistent ass,
And how dashingly he performs!

For five or six years, beginning, it seems, in 1952, we were in the habit of bringing in the New Year with a group of friends from the ICP, the Institute of Physical Problems, and the Lebedev Institute, and then on the following evening, January 1, we again gathered for the "day-after" holiday. The group rotated for several years around the Zel'dovich, Ryabinin, Landau, Ginzburg, Émanuél apartments, our apartment, the Semenovs', and many others. We made serious preparations for these festivities: there were presents, lotteries, and charades (in which YaB was particularly active). Once with Kompaneets we even wrote and performed (at the Landau apartment) a play "Day of a Scientific Man." These New Year parties, like the birthday of Kompaneets and Leĭpunskiĭ on January 4 and the day of the "Spring Stanyukalie" on March 3 (the birthday of Kirill P. Stanyukovich), became the times that YaB made regular trips to Moscow, and for us, in addition to the general merriment, they were days of joy at being with YaB, whom we all missed very much.

After 1956, YaB, by now a three-star hero, again made Moscow his main residence, although he did not yet return to the ICP, but joined Mstislav V. Keldysh at the Institute of Applied Mathematics. I had my neighbor back, and we began to exchange humorous rhymed messages, putting them in each other's mailbox. I was worried about being overweight, and regularly went in the morning in shorts to the Zel'dovich's apartment and used their large clinical scales to weigh myself. One morning I found the following missive in my mailbox:

Head Office of the Moscow Bureau for the Weighing of Live Commodities

Esteemed Clients Lyudmila Nikolaevna and Vitaliĭ Iosifovich!

The determinations, made at the night of the 5th to 6th of March, of the longitude, latitude, lunar phase, and acceleration of gravity have revealed a systematic error of -0.85 ± 0.07 kg in the weighing. All live weights marked on the belly before 6 March are incorrect and subject to change.

Head Astronomer of the HOMBWLC Ya. Zel'dovich

And here is an earlier message from YaB, after he had bought the poems of Simonov for me at the Academy's bookstore and, to my annoyance, had gotten this long-desired book thoroughly messed up with oil:

Bill

To His Excellency Mr. Financov—Gol—Donskoĭ

0. Delivery by personal car	20.00
1. Simonov	3
Oil on cover (butter)	1.40
2. Advance to Gastronom	56

3. Rustles not seen	100
4. Herring not taken	200
5. Princess of dollars:	
tickets	46
vodka 2 × 100	15
sandwiches	10
damage to morals	100
Total due	551.40

Here I could quote a great number of messages in verse and prose that we put in each other's mailboxes. They were quite diverse in genre, from very unrestrained (the first thing to hit the eyes is: "Mila! Do not read, I cannot be responsible for the decency of the contents") to extremely serious: "Vitya! Regarding the work of K. you would look at NO production in tracks of fission fragments specifically. The kinetics there is well known. There is an interesting technical problem: how to obtain NO thermally in a reactor. P., in particular, is working on this, but he works poorly. Don't you want that we enter together?" And after one of the jubilee evenings at the ICP some annoyance at his presentation: "What a lot of garbage I spewed out! Engels: labor made the monkey into a human being... I became a human being working at the ICP."

I keep every scrap of paper with YaB's bold handwriting on it as a treasured memory, but, of course, there is no need to make all of our serious and joking notes and messages known to everyone. I will restrict myself here to one more story, perhaps a particularly funny one.

In the mid-50s during the reconstruction of our Vorob'evskoye Shausse, it became urgently necessary to destroy some wooden, single-story huts near our house (Premier Aleksei N. Kosygin later appeared at the site of these huts). The destruction of these huts was a great disaster for the great number of bugs living there, and without delay they marched over to our house (unfortunately, I did not get to see with my own eyes their procession in close-ordered lines).

One very unpleasant night we could not get to sleep because of itching, and when we turned on the light we saw with horror that the walls were covered with bugs. In the morning we had to throw out the lower parts of our sofas, but it was impossible to do without the upper parts, because we could not sleep on the floor. After soaking these upper parts with kerosene, we put them on the landing of the stairs to air out. After two or three days small signs made by YaB appeared on the sofas. Arrows on which he depicted noxious insects were pointing to our door, and little "wrong way" signs were drawn behind the bugs, telling them that entry into the Zel'dovich apartment was forbidden. On the top of the drawing was written, "Bugs, *go home*."

At about this time YaB and I began to meet daily regarding science. We were interested in the properties of nuclei far from the ß-stability region. It began when Yakov Borisovich started looking for regularities in the energy of shell filling in light nuclei and arrived at the prediction that superheavy helium ^{8}He exists. I refined the predictions of the properties expected for this nucleus on the basis of the nonexistence of superheavy hydrogen ^{5}H. As is well known, YaB's prediction, which at first was met with skepticism, was soon brilliantly confirmed.

In comparing the different possible ways of estimating the properties of light nuclei, I encountered the work of Alfred I. Baz', which has opened the broad possibilities of various predictions based on the isotopic invariance of the nuclear force. YaB and I got in touch with Baz' and the three of us continued the work together, each in his own direction. YaB predicted the existence and properties of dozens of new neutron-excessive isotopes and analyzed the possibility of the existence of a neutron liquid. Baz' studied the dineutron problem in detail, and I worked on predicting the properties of many new neutron-deficient isotopes and the existence and properties of two-proton radioactivity. After combining all our results, in October 1960 the three of us published the review "*Undiscovered isotopes of light nuclei*" in the Soviet Physics *Uspekhi*, and 12 years later with Vladimir Z. Gol'dberg we published a monograph.

Thirty years have passed, but still I have a lively memory of the joy and instructiveness of daily scientific and human contact with YaB during those years, and I envy myself. Those periods of such enthusiasm for work, such impatience in the evening—if only the next morning would arrive more quickly, and with it new thoughts and new discussions, rare enough to count on the fingers. I am happy that one of those periods which I was granted was closely connected with YaB. I associate that period with favorite quotes of his: "I proved harmony using algebra" (Pushkin) and (addressed to anyone who pried into the details of his personal life) "I am a poet, which is what makes me interesting" (Mayakovsky) (and nothing else is your business).

However, science is science, and sport, especially skiing and all sorts of joking, always played an important part in YaB's life. I remember how we stayed in adjacent rooms in Uzkoe, and he and Varya (his wife Varvara Pavlovna Konstantinova, a woman of rare soul) raced ahead, hurtling down the hill, while I lagged some distance behind. On one of the ski slopes in the winter of 1957, when my other half Mila (then expecting our second son) had joined the company, YaB, watching my careful descent, said "watching you two, it looks like it is Vitya that is pregnant." In the evenings we had various medicines on the table next to the bed. One day, upon drinking up a dose of some ordinary prescription (valerian tea—Gol'danskiĭ), I choked from surprise because YaB had treated me to cognac, which was the same color. The large group of participants in the High Energy Physics Conference in Kiev in 1959 remembered the swimming competitions held in the Dnyeper river, which YaB organized between me and his eldest daughter Olya. I swim quite well, and so I couldn't at all understand why I was unable to get even a few strokes away from Olga and why everyone was laughing. When I was completely exhausted, I finally realized that YaB had had Olga wear flippers.

In 1963 we moved to another apartment in the same building, and our daily meetings ended. However, to compensate for this a new important topic of discussion arose: Academy affairs, because in 1962 I became a Corresponding Member. From time to time we arranged family outings to the cinema, and several times met, although we sat at different tables, at the New Year parties at the Central Writers' House. I was particularly proud of YaB's reaction to the report "Investigations in Gamma-Resonance Spectroscopy" which I read in February 1966 at the annual General Assembly of the Academy of Sciences. This report went very well, YaB liked it very much, and on the same day he suggested that we begin addressing each other with the familiar form of "you," ("ty" instead of "vy"). I do not conceal the fact that the possibility of addressing familiarly a man whom I worshiped as a living legend, especially in the presence of others, for example, my colleagues, flattered my self-esteem for a long time, until I finally got used to it.

The appearance of YaB at the rostrum of the General Assembly of the Academy of Sciences in December 1974 with a strongly-worded speech addressed to the heads of the Academy, which in the recent elections to the Academy had failed to give me a position, even though for the second time (the first was in 1968) I had received a majority of votes in my Division, provided some balm for my wounds.

As time went on heavy losses began to occur. In the summer of 1969 a close friend and brother-in-law of YaB, Academician Boris Pavlovich Konstantinov (about whom he wrote reminiscences) died suddenly. In 1970 Frank-Kamenetskiĭ, who had long been seriously ill, died. In 1974 we were literally stunned by the sudden death of Kompaneets, and in August 1976 Varvara Pavlovna died from a heart attack while vacationing in the Crimea. I remember how at night at Vnukovo we met the plane carrying her casket, how YaB went into the morgue with a single parting word. Not looking at anyone except Varya, staying away from everyone, he seemed to be telling himself the story of their first meeting and falling in love. The "twilight of the gods" was setting in.

However, ahead were still many signs of international recognition and trips abroad, first only to socialist countries, then to Greece and Italy, and finally to the United States in April 1987. YaB's talk on cosmology at the General Assembly of Leopoldina in Halle, GDR in April 1980 was a huge success.

He spoke in a quaint mixture of German and English, and in explaining the word "big bang" blew out his cheeks and then let out a loud "poufff!" Arriving in Washington in 1987 and going to the National Academy of Sciences, the first thing I saw was YaB at a table with drinks and sandwiches, taking a coffee break. Our meeting was unexpected for both of us and therefore all the more joyful. Of our subsequent meetings during this final year of YaB's life I remember a video evening at our house, in particular, watching the film "Bridge over the River Kwai" together, and, only two weeks before his death, our outing to the October Theater (Yakov Borisovich with his wife Inna Yur'evna Chernyakhovskaya and I with Mila) to see "Tomorrow Was the War."

Of course, I am leaving out many things; for instance, the story about the celebration of YaB's 70th birthday at the Institute of Physical Problems in March 1984 is worth describing in detail. However, I think that in this collection of reminiscences about YaB there are many other writers from the Institute of Space Research and the Institute of Physical Problems who want to tell about this jubilee and about the later years of YaB's scientific work. I am drawn to the earlier years, and for the last period of his life I now go directly to the time just before the end.

It happened that we saw and spoke with YaB on two evenings just before he died, November 29 and 30, 1987. Among the latest foreign scientific journals I had received were some containing articles about YaB and his work. One in *Science News* of May 30, 1987, described a humorous dialog between Zel'dovich and Gell-Mann at the meeting of the working group on quantum cosmology in Batavia during the first (and only) trip of YaB to the United States in April and May of 1987. The issue of *Nature* of October 29, 1987 was specially devoted to Soviet science, and an article entitled "Great Men and Barons" was devoted to YaB. Finally, *Chemical and Engineering News* of August 31, 1987 contained an excellent review by George Fisk and James Miller on the chemistry of combustion. When he heard about this, YaB ran over to leaf through them, and on the following day he took these journals to read. He was particularly interested in the review on chemical combustion, which gave him deep satisfaction because it clearly showed that after 40 years his classic work on nitrogen formation in combustion, on the combustion mechanism itself and flame propagation, on chemical reactions in flames, and on detonation remained "evergreen."

We did not see each other on December 1; I only learned from my son that YaB had asked to be given a lift to the Institute of Physical Problems, and this meant that he didn't feel well. At midnight on December 2 we heard the sounds of a car in the courtyard and noise on the adjacent staircase; Mila ran to find out what was going on and reported that YaB had been rushed to the hospital. I then phoned the now-deceased Igor' K. Shkhvatsabaya to take special care of Yakov Borisovich. We were filled with alarm, but still far from feeling close to the tragic outcome hanging over us. He had only fourteen hours left... .

A few days after the death of Yakov Borisovich I learned that Rudolf Mössbauer wanted to nominate him for the Nobel prize and was undecided only in what field to nominate him: chemistry or physics.

I also heard many other eminent scientists from Western Europe and the United States say they believed that Zel'dovich highly deserved to be a candidate for the Nobel prize. Do I think that this amazing versatility complicated the classification of him as a physicist or chemist? Actually, in the novelty and brilliance of his ideas and the value of his contributions to science, Yakov Borisovich was certainly a scientist of Nobel caliber both in physics and in chemistry.

The further the day of our parting recedes, the stronger the feeling of the loss to science. As far as we his friends are concerned, he will remain with us until the end of our own days. We simply cannot get used to and make peace with the idea that he is no longer with us.

Note: See also author's obituary of Ya. B. Zel'dovich in Physics Today **41** (12), 98–102 (December, 1988).

SCIENCE, SOCIETY, AND HUMANITY

I should confess that the second section of Part I is somewhat eclectic.

The earliest article here seems to be of some interest as retrospective—at that time (1968) computerizing, informatics, and scientometry were still underdeveloped in our country.

Two selections of "preperestroika" period (1984–85) have an autobiographical shade, while later publications reflect the preoccupation with general and specific difficulties and weaknesses in the development of science in the USSR. In May 1989 I was elected as a Member of Soviet Parliament—Congress of People's Deputies of the USSR—and soon became the Chairman of Academical Parliamentarians Group.

The article "Soviet scientists face perestroika's challenge" was composed as a program of views, demands and actions of this group.

What Should a Scientific Journal Be?

Among the stupifying numbers illustrating the growth of science in our times, one that stands out is the number of scientific journals. According to UNESCO data, in 1800 there were only 100 scientific journals published in the entire world, while in 1950 there were nearly 100,000, and today over 200,000.

The simplest analysis of present data on scientific periodicals leads to staggering conclusions. Indeed, if the total number of scientific (monthly) journals is 200,000—their average circulation 1000 copies, average size 100 pages—and if the users of every library read no more than approximately 10% of the articles published in each issue of any journal (of course, different pages in different libraries), that would mean that all over the world approximately 20 billion pages of scientific journals every month remain virgin clean, without even a single reader.

It is becoming more and more difficult to sort out the flood of scientific articles, the main source of information for scientific researchers. This accounts for the number and particularly the size of the various collections of abstracts and bibliographical indices. A greater and greater role is being played by the publication of reviews which summarize the main results of wide-ranging research.

However, in spite of the extremely fast growth in the number of scientific journals, the amount of scientific information is not growing. Moreover, the time to publish an article, that is, the time between submission of the article to the editor of the journal and its publication, is as a rule increasing. And this is occurring in spite of the general trend for the length of scientific articles to decrease.

For example, the average time for publication of articles in the *Journal of the Russian Physico-Chemical Society* in 1916 was two and a half months, while for the *Soviet Journal of Physical Chemistry* in 1966 it was 19 months for articles and 12 months for brief reports. And this is occurring at the same time as science has begun to develop so rapidly that the danger of an article becoming outdated while it lies on the editor's desk quite often becomes reality.

What can we do? Is it really necessary to further increase the number of scientific journals without changing the form of publication itself? It is thought that this solution would be incorrect. In order to justify the rationality of other solutions, let us first note one very important fact. Given the enormous growth in the number of scientists and the amount of their published output, there is one indicator which has hardly increased, and has possibly even decreased, during recent decades: the average number of readers of each scientific article.

The actual number of such readers can at present be judged mostly indirectly—from requests for reprints received by authors, from references to a particular article, and finally from the demand for so-called "preprints," produced without typesetting. These preprints are now published by many large institutes here and abroad, in order to avoid the months-long delay in the journal publication of the article, and to inform all specialists in a given topic (who are usually well known to the authors of the article) as early as possible about new results. The number of preprints produced usually does not exceed 100 or 200 copies. It is this number which should be considered meaningful. It can be stated that the average number of readers interested in the full text of any scientific article is a few dozen, and in rare cases a few hundred. If we also take into account the fact that the readers of scientific journals are grouped according to institute and use libraries, it becomes clear that the needed number of copies of the full text of scientific articles published in journals does not exceed several hundred. Even several dozen copies are often sufficient.

As a rule, a scientific researcher manages to read in detail only a very narrow circle of articles, since the number of publications grows so rapidly. A scientist in the best case becomes acquainted with the problems in other areas from reviews or from brief "Letters to the Editor," or, finally, from the short abstract preceding the text of an article.

It therefore seems necessary to gradually cease altogether the publication in scientific journals of what today is their main content: the full texts of scientific articles. Instead, journals should print reviews, brief reports (or letters to the editor) about the most important new results, short author's abstracts—notes describing the content of articles submitted to the journal. The editor or publisher or a special institute for scientific and technical information should send the articles themselves only by request, often upon payment. The articles can be published without typesetting, like preprints.

It will not be difficult to determine the corresponding payment not only for copies of an individual article, but for other possible requests such as an annual delivery of all articles of a particular author or all publications on a particular problem.

It seems that this route of development of scientific journals promises a number of advantages, of which saving paper will not be the sole or even the most important one. The time for publication of author's abstracts of articles could rapidly be reduced to two or three months while preserving the present number and sizes of journals (and perhaps even reducing them). The accounting of the prevalence and dynamics of development of research on a particular problem and in an entire field would be simplified. Journals would become points of contact between scientists of different cities and even countries working on similar problems. Reader requests for articles becomes a simple and objective (not error free, but on the whole very indicative) criterion of their value.

For several decades now our journals have been translated cover-to-cover into English and published abroad. The price of the translations is much higher than that of the originals. For example, a single issue of the journal *High-Energy Chemistry* costs 75 kopeĭkas in the USSR, while the translation costs 30 dollars (the price of a single translated article is 15 dollars). In our proposed scheme we could translate the full text of an article (retained by the editor) ourselves and then supply copies abroad by request, thus receiving a sizeable income.

The forms of publication listed above are not at all novel. For example, the "Letters to the Editor" section in our very popular *Journal of Experimental and Theoretical Physics* and "Brief Reports" in the widely circulated chemical journal *High-Molecular Compounds* are published quite rapidly. The analogous special edition of letters to the editor of the most important American physical journal *Physical Review Letters* also contains short abstracts of articles submitted to the journal, which are published many months before the articles themselves. Because of this readers can often request and obtain the full text of a journal article from the authors even before it is published.

The All-Union Institute of Scientific and Technical Information (VINITI) takes care of the acceptance, storage, and circulation of non-typeset manuscripts in the natural, exact, and technical sciences. Readers can obtain copies of articles quite cheaply: the price of a single page is only 0.7 kopeĭka (for a printing of over 300 copies) to 1.8 kopeĭkas (for a printing of less than 200 copies). The journal *Atomic Energy* also keeps some manuscripts, and provides copies upon special request. However, at present the fraction of manuscripts circulated in this way is negligible, and the act itself of keeping an article instead of publishing it is often viewed as a sort of discrimination.

Meanwhile, the problem is to make the storage of the manuscript of an article and its circulation only upon special request the rule rather than the exception.

The changeover to a new system of producing scientific journals is not a simple matter. It requires changing many old procedures and acquiring new habits and new ways of operation. In addition, it is not possible to recommend a universal solution applicable to all journals: a special case might be made, for example, for medical journals or for journals like *Zavodskaya Laboratoriya [Industrial Laboratory]*, which have wider circulation and a broader range of topics. (Of course, none of this applies to journals published for mass readership.) It would be reasonable to test the new system for one or two journals, ones with an "excess" of articles which have been lying on the editor's desk for years. I am confident that the experience with these journals will stimulate many other publishers to follow their example.

Note from author: This article was published by *Izvestiya* in 1968. At that time I was still unfamiliar with the extremely important and useful practice of the Institute for Scientific Information in Philadelphia. This and similar practices and the wide computerization of the storage and dissemination of scientific information have created many new factors in the problem of scientific periodicals. Nevertheless I decided not to exclude the 1968 newspaper article from this collection of essays—some thoughts expressed there still seem to be topical.

Not Only about Chemistry[a]

[Editor's Note: The past and future of chemistry, its theoretical and applied problems, its present role among other disciplines, essentially, its role in our lives. This is the topic of conversation of the head of the Section on the Structure of Matter at the Institute of Chemical Physics, USSR Academy of Sciences, Vice-Chairman of the board of directors of the All-Union Society "Znanie," Lenin Prize winner, Academician Vitaliĭ I. Gol'danskiĭ, and our correspondent Marina Kuryachaya.]

Kuryachaya: *Vitaliĭ Iosifovich, our conversation today should concern not only current problems in chemistry, but also its past. Throughout history and at the present time there have been interesting connections between chemistry and other areas of knowledge, in particular, physics and biology. On the one hand, specialists today are witnessing the spread of ideas from chemistry to medicine, biology, and other sciences; a sort of "chemicalization" of thought is observed in them. On the other hand, chemistry is considered, mainly by physicists, as a subject which is not original, which does not have its own theory. Chemistry, in their opinion, is a code of empirical rules which can be justified theoretically only by physics. Which side to you think is correct?*

Gol'danskiĭ: First of all, I would like to say a few words in defense of empirical rules. Very often they are a necessary stage in the construction of a theory. The periodic table of Mendeleev was also empirical, but very many valuable physical discoveries were made using it. It was essentially the periodic law together with radioactivity which provided the foundation for the present ideas about the structure of matter.

Now a word about originality. Chemistry does not have indeed its own fundamental theory. And, of course, any theory which it obtains in the end comes from theoretical physics, a science applicable to all the phenomena of nature. The structure of matter, in general, the nature of the microworld, is

[a] Interview by Marina Kuryachaya.

essentially described in the language of quantum mechanics. But to therefore speak of a subject as not being original would be a clear hurt to chemistry.

If we argue in this way, we can also accuse physics of not being a fundamental science, because the relations between the physical characteristics of matter are written in the language of mathematics. Incidentally, chemistry also has its own language: the symbols for the chemical elements. And the relations between chemical characteristics—reaction equations—are written in the language of chemistry. In this sense chemistry is even in a better position than physics.

But another aspect of the problem is hidden in your question. I cannot find an accurate name for it.... Let us refer to it as "degree to which something is fundamental." Let us consider the hierarchy: elementary particles, nuclei, atoms, molecules, and all possible supermolecular structures. In this hierarchy chemistry, strictly speaking, governs only the level of molecules. Preceding levels are governed by physics, and subsequent, the supermolecular levels, are governed either by biology or again by physics. In a sense, chemistry is just squeezed in.

But in fact the world of molecules is so large that chemistry possesses an enormous stock of knowledge.

However, if the degree to which a science is fundamental is characterized by a fraction with the theoretical foundations of the science as its numerator and the applications as its denominator, then it can really seem that chemistry today is a less fundamental science than biology and physics. But this is not because the numerator is small, it is because of the enormous number of applications of chemistry. This fact should be kept in mind.

Clearly, from a fundamental viewpoint, chemistry has a long and great future ahead of it.

Kuryachaya: *Can the absence of a fundamental theory in chemistry be explained by the fact that chemical substances are more complicated than physical ones and, accordingly, the problems faced by theoreticians are more complicated?*

Gol'danskiĭ: Well, no, I do not think it can be stated this directly.... Judge for yourself which is more complicated: a solid object or an individual molecule making up the object? The solid object contains many molecules. Can it be maintained that the level of its structure is much more complicated? In some sense yes. But in return the phenomenology, the description, is simpler precisely because it is made up of many molecules: we can neglect some details and at the same time introduce some new concepts into its description.

It is therefore impossible to say that the theory of molecules is more complicated than the theory of the atom, and that the theory of a solid material is more complicated than the theory of a molecule. In going to each new level we find it possible to describe phenomena and we lose something. It is as if we increase the complexity and at the same time overcome it.

For example, today the leading edge of science deals with what would appear to be the simplest objects, elementary particles. But so far they are not well understood. Completely new ideas and new quantum characteristics of matter arise; by this is meant the unified field theory, in which elementary particles must enter as various manifestations of a single field. This theory combines interactions of different types: the electromagnetic and weak and strong interactions, and, perhaps in the future, the gravitational interaction, too. So far the definitive version of such a theory has not been constructed.

Of course, it cannot be said that since physics deals with simpler objects, chemistry with more complicated ones, and biology with even more complicated ones, then physics is simpler than chemistry and chemistry is simpler than biology... .

Kuryachaya: *You mentioned the discovery of the periodic law. This is a vivid example of how the solution of a problem important at one time gave a powerful impulse to new ideas and directions in science. What about today? In which area of chemistry do you think research is most important from the viewpoint of fruitful new ideas, experiments, and their influence on scientific thought?*

Gol'danskiĭ: To answer this question we need a general view of science, of all of natural science as a whole. Three very important points of growth can be seen here. Two are in physics and one is in biology. In physics they are elementary particles and cosmology, that is, questions of the origin and development of the universe. And in biology is the most important problem: the origin of life and its evolution. The most important area in chemistry in the future would seem to be chemical bionics (in other words, biomimetics), the chemical modeling of systems produced by living nature. Let me explain what I mean.

Biological systems are record-holders: first, because they have the largest capacity for information, cf., the human brain. Second, consider substance fluxes: the production of substances in a living cell is larger than in any factory (per unit of working weight). Finally, consider energy fluxes: in bioenergetics the productivity is higher than in any other system, and here we see the highest efficiency, for example, in photosynthesis. It is remarkable that the transformation of chemical energy into mechanical energy occurs practically at constant temperature.

But today all this comes only from nature. If only systems possessing the three parameters I listed at the level of biological systems could be created artificially... . Learning the principles of the operation of living systems and being able to reproduce them by means of chemistry is what chemical bionics is. The problem is so fundamental that in future it can ensure that chemistry plays a very important, I do not want to say leading, but a very important role in both applied and theoretical problems.

One is the creation of artificial enzymes—catalyzers, which would have the same level of selectivity and activity as biological ones. Or artificial membranes, which could work like biological ones: allowing some ions to pass through and not others, possessing one-way permeability.

The creation of such chemical models would be interesting not only from the practical point of view. It would also shed light on how processes occur in a living cell.

Kuryachaya: *In other words, this would make it possible to some degree to approach the problem of the origin of life?*

Gol'danskiĭ: That question is special.[1] But since here the approach is physical, that is, the origin of life is considered as a physical problem, which does not reduce only to chemistry, I will not attempt to discuss it.

Kuryachaya: *But this is the most general problem of natural science, isn't it? And it is particularly interesting. Therefore, could you please at least say something briefly about it.*

Gol'danskiĭ: The question is posed as follows. Can the appearance of life be reduced to evolution? Or is it not a smooth evolutionary process, but instead some sort of "catastrophe," a phase transition?

For example, in cosmology there is a theory of the origin of the universe, the so-called Big Bang theory. Did something similar to such an "explosion" occur in the development of organic nature causing a transition to bioorganic nature, to the appearance of life? It is well known that in nature there exist only left-handed amino acids and only right-handed sugars, and this is one of the most important properties of living systems distinguishing them from nonliving ones. How did this breaking of mirror symmetry occur?

There are essentially three factors which need to be taken into account. The first is purely fluctuational, statistical: suddenly in some place in a homogeneous mixture by chance the number of left-handed molecules (compared to right-handed ones) exceeds the number of their right-handed antipodes.

Second, there is the advantage factor, which operates in such a way that this inequality is no more of accidental origin. Such might be e.g. the neutral

[1]In recent years I had a chance to think and work a great deal on this problem with an extremely talented scientist, Leonid Leonidovich Morozov, a doctor of physico-mathematical sciences. I am strongly convinced that he would have had a great future in science, but he died suddenly on June 9, 1984, two days before his 38th birthday.

currents known in elementary-particle physics; they are one of manifestations of the so-called weak interactions. Because of these neutral currents we can get the following picture: the energies of left-rotating and right-rotating molecules differ by a very tiny amount. Of course, the advantage factor must operate for a very long time. Say that for billions of years a test, an experiment, is repeated and repeated, and each time there is a small advantage in reactivity for one of mirror antipodes. But could this gradually, in an evolutionary manner, eventually cause left-handed amino acids to survive, while right-handed ones gradually turn into left-handed ones? And thus break the mirror symmetry of the world?

Finally, the third factor. It operates in the opposite direction: the so-called racemization factor. Essentially here the existing left-handed and right-handed forms, interacting, extinguish each other, transforming into a neutral medium, as though they had annihilated.

The interplay of these three factors could cause one mirror form to become dominant over the other in nature. However, can this be explained as an evolutionary process, a gradual change? Or should we view it as an "explosion," a transition through some critical condition? Some changes occur, they accumulate, and all of a sudden the system becomes unstable. From being nearly symmetric it changes promptly into being purely asymmetric.

Evolution or a sort of phase transition, the explanation of what role neutral currents play in this, the attempt to observe their contribution to chemical processes, this is the absorbing problem. And if we speak of personal biases, to me at present it is the origin of life that is the most interesting problem.

Kuryachaya: *You mention personal biases. In connection with this, the following question: you have managed to accomplish a great deal in science; what work is especially dear to you and why?*

Gol'danskiĭ: You know, I would ask that question somewhat differently. When I think of what I have done, I look back, remember, and then think: which years were the most successful? Which stage of past works now appears to have been the most fruitful? And I get the impression that they were the years from 1959 to about 1961. During these two or three years the foundations were laid for essentially all the main directions of research which I later did and which later became especially dear to me and produced important results... .

In 1959 we were carrying out experiments at the synchrotron of the Lebedev Physical Institute. It turned out that the characteristics we found for gamma-quantum scattering on the proton showed that the latter could not be viewed as a point particle. In our experiments it was possible to conclude that the proton is a complicated particle possessing a polarizability. Moreover, the electric polarizability constant was determined numerically. That work (it was mentioned in textbooks and in the encyclopedia *Physics of the Microworld*) later became the subject of an officially recognized discovery. The laborious, time-consuming experiments had a gratifying outcome. Their results were reported the same year in Kiev, at the High Energy Physics Conference. That was the first result of the year.

Here, probably, I should explain that in 1952 I moved from the Institute of Chemical Physics to the Lebedev Institute, worked there seven years, and then in 1959 began to gradually prepare to return to chemical physics, in order to combine the experience I had in chemistry with that I had in nuclear physics. Formally, this move occurred only in 1961. But actually, beginning in 1959, I combined my work at the Lebedev Institute and at the Institute of Chemical Physics.

The experimental work at the Lebedev Institute proceeded as planned. Meanwhile I took a lively interest in the contribution of tunneling transitions to the chemical reaction rate. I made some theoretical calculations for myself and arrived at the conclusion that at certain low "cross-over" temperature the contribution of quantum tunneling transitions through the potential barrier between initial and final states should exceed the contribution of classical over-barrier transitions and at further reduction of temperature the reaction rate must finally emerge onto a plateau. In other words, it should exist a low-temperature limit of the rate of exothermic chemical reactions.

I reported these results also in 1959 at the Mendeleev Congress which was held in Moscow. There I met the distinguished British scientist Ronald Bell, one of the pioneers in the study of tunneling in chemistry, and he paid deep attention to my work. Therefore, in 1959 it was determined the second direction in my interests. The research in this area became principal for us later, in the late 1960s and 1970s.

Next came the problem of the properties of nuclei far from the beta-stability region, in particular, nuclei with a strong neutron deficiency. I became interested in this also in 1959 and also "for myself." And I found a way to predict the properties of such nuclei. I derived formulas and used them to calculate the expected masses of many unknown isotopes and saw that two-proton radioactivity must exist. Therefore, the result, which after a long time is now beginning to be confirmed experimentally, was obtained in 1960. This certainly was one of my favorite studies.

And again, in connection with my move from the Lebedev Institute back to the Institute of Chemical Physics, in 1960 we did our first work on the positronium reactions in aqueous solutions. This was experimental research, which became the first in a very large number of our studies on the chemistry of the positron and positronium.

Finally, in 1961 we began working on chemical Mössbauer spectroscopy.

Our first experiments were carried out with colleagues from the Nuclear Physics Institute of Moscow State University. The Mössbauer effect was discovered in 1958, and in 1961 we observed this effect in tin-containing polymers. In 1962 we found and explained the asymmetry of the Mössbauer spectra; this later became known as the Gol'danskiĭ-Karyagin effect.

In a word, this period was the most fruitful. I recall it with great satisfaction. I do not mention the fact that I was twenty-five years younger, an important advantage... .

Kuryachaya: *Vitaliĭ Iosifovich, sometimes not only achievements but also errors have a great influence on the development of science. In the vain searches for the philosopher's stone and the elixir of longevity, alchemists accumulated an enormous amount of useful information which formed the basis of the future chemistry. In attempting to separate the nonexistent phlogiston, chemists discovered the secret of combustion and discovered oxygen. In short, despite the certainly mistaken direction of their research, important discoveries were made and fruitful ideas arose. Do examples like this exist today?*

Gol'danskiĭ: An example of an error with great consequences? In my opinion, a great error is generally a privilege of a great human. And here I am involuntarily reminded of the mistaken discovery of the transuranium elements by Enrico Fermi.

As is well known, he studied the interaction of slow neutrons with various nuclei. And he found that neutron absorption and nuclear transformations occur with great efficiency here. When Fermi reached uranium, various new isotopes began to be obtained. At first it was assumed that the elements 93, 94, 95, 96, and, I think, 97 beyond uranium had been discovered. This was announced. It later turned out that Fermi had not observed the transuranium elements. The error tormented him to the end of his days.

However, less than 10 years later, after more research had been done in this area, it became clear that neutron capture by uranium actually leads to the production of neptunium, and then neptunium is transformed into plutonium and so on, so that the transuranium elements are obtained. But this happened later.

And the error of the great man in the end became the source of a great discovery. It signaled the beginning of the atomic era. Many researchers followed in Fermi's tracks and began to seek what he, in the final analysis, had observed... . And soon after Pavle Savić and Irene Curie carried out their remarkable work, Otto Hahn and Fritz Strassmann discovered fission... .

When you spoke of errors, you touched upon a serious topic. I would like to digress a bit and stress another important point.

Every scientist has the right to make a mistake. But then often when one is made some bystanders try to crush the person... . I think that this is not correct behavior in science.

It is possible and even necessary to crush someone if an attempt is made to cover up, to paint over the error instead of recognizing and correcting it. And, conversely, whether it is a gross error or a small one, pointing it out is the same as stopping and warning others after running into a hole in the road. This is high-minded. But what if—I continue the analogy—one tries to hide the hole and drive farther, so that others then run the risk of running into it? This is completely unacceptable from the viewpoint of morals, scientific ethics, whatever... .

It is very important that errors become public property, that the author himself and other scientists work persistently to subvert them, to understand the reasons for them. It would be quite useful to see a book entitled *The History of Great Errors*. It seems to me that this is a very interesting problem for the Institute of the History of the Natural Sciences and Technology of the USSR Academy of Sciences and for authors working in the field of popular scientific literature.

And there is something else I would like to mention: it is not necessary to fear errors. Dangers of this type, the fear of being laughed at, can halt very important research. It is sometimes necessary to be very brave in order to continue working. But this is essential.

I have been told that, in his time, many laughed at Pavel Alekseevich Čerenkov: they accused him of catching devils and seeing some sort of rays in the darkness, which no one but he could see. And Čerenkov, in spite of the hecklers, stubbornly pursued his goal. In a surprising manner he essentially not only observed this radiation with the naked eye, but also managed to ascertain some of its fundamental properties. Čerenkov's work became the subject of quantitative study and became one of the most important achievements in our native and in international science, and was crowned with the Nobel Prize.

Kuryachaya: *No fear of errors, ridicule; the stubborn pursuit of one's goal—these are apparently always necessary for a discoverer of something new and, indeed, aren't they particularly important for defending unusual, "crazy" ideas?*

Gol'danskiĭ: You know, at times it seems to me that today there is a sort of careening in people's consciousness. Here it is very important to stress the fact that in the great majority of cases "crazy ideas" are really crazy, without quotation marks. A remarkable phrase about "crazy ideas" is often used to say "and why not?" They say, Einstein could, so why can't we? This is a sort of *a priori* justification of completely idiotic pseudo-discoveries, which only take up a great deal of time to see that they certainly contradict science.

It should be born in mind that a new theory, as a rule, extends the explanation of phenomena to some new region, and at the same time coincides with the old one and encompasses its knowledge. If the idea put forward contradicts established facts, then it is truly crazy. This needs to be understood well, so that there is no careless speculation or claims that a crazy work and a correct work are the same... .

Kuryachaya: *Vitaliĭ Iosifovich, our conversation is nearing its end. I would like to conclude with a somewhat traditional question. How do you see chemistry in the year 2000?*

Gol'danskiĭ: We have now gotten into the habit of predicting and, of course, the most tempting date is the year 2000. Before answering, I would advise you to take a look today at old journals and see how correct were the predictions made for the fifties, sixties, and seventies. You will see what a thankless job it is to predict something.

It seems to me that it would be a very good idea to publish studies analyzing in which cases a prognosis proved too modest, and in which it was unrealistic. This is very important, both from the viewpoint of analyzing the past and from the viewpoint of extrapolating to the future. In my opinion this would be very instructive.

Now with this reservation I will answer.

Chemistry in the year 2000... . There is, of course, the problem of creating artificial enzymes, artificial membranes, artificial energy transformers, artificial information devices. I spoke earlier on the theoretical aspect of this problem. Now I will speak on the applied aspect, although, of course, "applied" is a rather arbitrary term.

Here it is suitable to recall the works of Mark E. Vol'pin and Aleksander E. Shilov on nitrogen fixation using a sort of artificial enzymes. The development of such soft catalytic methods can lead to the utilization of the enormous nitrogen reserve surrounding us and to the creation of valuable products of a different type.

A membrane is essentially a special kind of chemical semiconductor. It is capable of transport even against a gradient, that is, the agent goes from the region of lowest concentration to the region where its concentration is highest. In other words, active transport of substance is possible.

Energy transformers... . For example, artificial muscles, in which chemical energy could be transformed into mechanical energy with high efficiency.

Information devices... . The creation of artificial neurons—chemical models of those information elements which in the final analysis give the enormous information-processing capability of our brain. If we compare the brain to the most perfect integrated circuit, the limit of the information capacity of the brain, its tiniest element is the molecule. Imagine a molecule performing the functions of a transistor. This "transistor" is extremely small and, accordingly, contains an extremely large amount of information per unit volume. Ideas of creating molecular circuits of this type in modern electronics—essentially molecular electronics—are already being discussed in the press.

A bit about chemistry as applied to problems in genetic engineering. There are remarkable studies by our famous geneticist Iosif A. Rapoport, which were awarded the Lenin prize in 1984: the cycle "The phenomenon of chemical mutagenesis and its genetic study." Basically, they describe how chemical materials can be used to induce mutations, which are extremely important for selection, for deriving new species. And it is very interesting to push forward from the discovery of chemical mutagenesis to the study of it at the molecular level: how to act on a molecule, inducing some *a priori* specified mutation into it; this is an extremely interesting problem.

Now, chemistry as applied to medicine: here there is a big assortment of problems for chemistry. The creation of artificial tissues and organs, the use of which would not lead to tissue rejection or induce the formation of blood clots. Very promising work in this area has been done by our Nikolai A. Platé. Then there are problems of immunology, artificial vaccines, which Rem V. Petrov and Victor A. Kabanov are working on. Or, for example, cancer chemotherapy, a very serious problem... . It would be interesting for you to report on the achievements in this area of Nikolai M. Émanuel.

A few more words about the problems of chemistry as a science. It is very important to study more and more rapid chemical processes; there are many new things here in the understanding of elementary chemical transformation events.

Of course, chemistry must play an important role in solving many physical problems. For example, in the search for high-temperature superconductors. Today this problem is to a large extent a chemical one, not a physics one. Or the creation of an entire series of new materials, say, polymer films based on polyacetylene possessing conductivity.

Chemical methods are also being applied in fundamental physics. An example is the problem of the neutrino-induced transformation of one element into another. Here the problem consists of the radiochemical separation of the products of these reactions and the determination of the neutrino flux.

And, returning to conventional chemical problems, I would mention the intensively developing field of laser chemistry, namely, the problem of selective action on chemical bonds, using extremely fine tuning of the wavelength of a laser beam, to learn to control very complicated chemical processes. This might be of both theoretical value and importance for chemical industry, and chemical mutagenesis and, in the end, for genetic engineering... .

I have already mentioned many things, and there are more. It is impossible to comprehend the immensity, but also impossible to remain silent about the extremely important role which chemistry has been called upon to play in the food- and energy-production programs.

In the food-production program there are the problems of artificial food, fertilizers (nitrogen-fixation again), herbicides and pesticides, the search for optimal methods of reprocessing raw materials for food. In connection with the energy-production program there are the problems of finding the most efficient methods of capturing solar energy, harnessing such energy-carriers as hydrogen and carbon monoxide, gasification and coal liquification (the complex use of radiation and high-temperature heat from reactors of nuclear power stations may prove to be useful for these purposes). A very important problem is reprocessing, repeated utilization of raw materials; if we learn to reuse them we can save an enormous amount of energy... .

I see that it is time to state some conclusions. They are very simple. Today chemical experiments are interesting not only for chemistry, but also for physics, for biology, and for other areas of knowledge. And chemistry with its methods and the range of problems which it studies still has ahead of it a great future in the modern natural sciences.

I Trust in the Human Mind[a]

The most surprising event in the never-ending attempt to understand the mysteries of the universe is an unexpected discovery, when what yesterday seemed to be an indisputable truth today becomes only a special case, a single link in the chain of knowledge.

Let me give an example. One of the fundamental laws of chemistry which each of us encounters on a daily basis when putting food in the refrigerator is that the chemical reaction rate slows down when the temperature is lowered, and near absolute zero all chemical activity must cease altogether. Only relatively recently has it been possible to demonstrate experimentally that even at superlow temperatures, say, minus 260–270 degrees Celsius, polymerization processes occur millions of millions of times more quickly than predicted by the laws of classical chemistry. This discovery made it possible to take a new look at prebiological evolution, a sort of "cold prehistory of life." It is quite probable that processes leading to the formation of organic molecules, including proteins, can occur in the low-temperature, lifeless cosmos.

I spoke with the author of this and many other amazing discoveries in chemistry and physics, Academician Vitaliĭ Iosifovich Gol'danskiĭ, Lenin Prize Laureate, Chairman of the Scientific Council of the USSR Academy of Sciences on High-Energy Chemistry and of the Commission of the Academy of Sciences on Synchrotron Radiation, and vice-Chairman of the All-Union Society Znanie. (Knowledge)

Back in the early 1960s V. I. Gol'danskiĭ discovered a fifth type of radioactivity—two-proton radioactivity—"at the tip of a pen." This discovery found first experimental confirmations (in the United States) only recently, nearly a quarter of a century later.

[a] Interview by Evgenii Dugin.

Dugin: *Vitaliĭ Iosifovich, is the route from theory to experiment and practical application always so long?*

Gol'danskiĭ: Unfortunately, the delay between theoretical predictions in science and their practical confirmation is often very long. Here a lot depends on the requirements of the age. If an idea is consistent with the laws of nature, even ones not yet discovered, then mankind will need it in a decade or a century. For the route to practical exploitation to be short, the idea must, as they say, be in the air. In the case we are discussing, the delay between the prediction and the experiment which gave first evidence of the emission of two protons amounted to a bit more than two decades. This is long enough on the scale of a human lifetime, but only an instant on the scale at which knowledge is acquired. Such cases are not uncommon in science.

Dugin: *Could you discuss in more detail this discovery?*

Gol'danskiĭ: In the past it was assumed that there are only four types of radioactivity: alpha decay, beta decay, spontaneous nuclear fission, and proton radioactivity. However, an analysis I carried out in the late 1950s of certain properties of nuclei indicated the possibility, rather, the necessity, of the existence of yet another type of decay. This is the decay of nuclei in which they emit not simply protons, but a pair of their constituents (nucleons): protons or neutrons. Analysis of the properties of nuclei showed that there must exist some isotopes of even elements in the Periodic Table possessing two-proton radioactivity.

The point is that in nuclei of even elements there occurs a so-called pairing of protons with opposite spins, or, to put it in simple terms, protons rotating about their axis in opposite directions. This process is accompanied by the release of a rather large amount of energy, millions of times greater than that associated with the pairing of electrons in atoms and molecules. In spite of this difference in the energy, the physical nature of two-proton radioactivity is similar to the formation of chemical bonds between atoms and molecules or to the appearance of electron pairs in metals, which gives rise to superconductivity at low temperatures. All these phenomena originate in the tendency for electrons, protons, neutrons, and other such particles to form pairs.

Dugin: *What, in your opinion, delays the practical implementation of an idea?*

Gol'danskiĭ: In addition to objective factors such as insufficiently well-developed experimental techniques, there are also important subjective factors which sometimes prove decisive. Often the unwillingness of people working in one field of specialization to learn what is going on in related fields slows down scientific-technological progress as a whole.

Let me give an example of how important it is to be familiar with fields different from one's own. The theory of one of the most remarkable discoveries in nuclear physics in our times, the "Mössbauer effect," was known for more than 20 years before the discovery of the effect, but in a related area, for neutrons rather than for gamma quanta. And for long decades people working literally two steps away from a great achievement managed not to notice it. You can imagine how many specialists tore their hair after reading the first reports of Mössbauer.

Dugin: *In transforming the accepted ideas, the fundamental sciences sooner or later lead to practical results. What if the need for the results of a discovery is already felt by society? How in such a case does science serve practical goals?*

Gol'danskiĭ: As is well known, the main task of academic institutions is to develop the fundamental areas of science. However, deep satisfaction always comes from the realization that one's work is providing some definite benefit to society. Let me give an example from my institute, in which I have spent nearly all of my working life. In the laboratories of the Institute of Chemical Physics many cases of the creative action of an explosion have been studied. Instead of the usual destruction of matter, it has proved possible to use a shock wave to transform graphite into diamond, to synthesize a number of materials with valuable properties, to effect instantaneous polymerization, and to transform raw rubber into vulcanized rubber. A wide range of modern physical techniques is used to realize new effective chemical processes, for producing specially designed new materials with specified properties.

Dugin: *Vitaliĭ Iosifovich, you will agree that scientific schools play a large role in discoveries. What are the features of the Gol'danskiĭ school?*

Gol'danskiĭ: I do not like such words as applied to myself. The sort of group in which I work well is a group of people with the same views as my own. We have been working together for a long time. Of course, we have developed our own style of dealings and mutual understanding. Perhaps this can in some sense be described as a school.

However, defining a scientific school is not so simple. For example, my friends and colleagues at the institute can justifiably claim to belong to the school of Academician Nikolaĭ Nikolaevich Semenov, although many of us were not really his students. Academician Vladislav Vladislavovich Voevodskiĭ, who was one of the most prominent representatives of this school, was a student of Academician Viktor Nikolaevich Kondrat'ev, who, in turn, was one of the first disciples of Semenov. Meanwhile, in speaking of the "Semenovites" we should not forget that Semenov himself is a brilliant representative of the famous Physico-Technical School of Abram Fedorovich Ioffe.

It is difficult to say where one school ends and another begins. It is like defining the sources of Russian literature. Dostoevskiĭ said that it all originates from Gogol's *Overcoat [Shinel']*. The same can be said in defining scientific schools... .

Dugin: *Anyhow, without continuity there is no such thing as a scientific school. What are your relations with young people? Are you interested in the relation between students and your own scientific work?*

Gol'danskiĭ: You realize that the answer to this question can involve lots of trivialities, such as "I have excellent relations with young people," "I look upon students as my descendants, as my future," and so on. All this is both true and false. You probably cannot find any scientist who is indifferent to the opinion of his students regarding himself. You want them to be satisfied with their interrelations with you.

In principle I do not at all like the word "scientist" ("learned"). I remember how Landau used to laugh at it: "A learned cat I can understand, but what on earth is a generally learned"? Instead of "scientist" I prefer the term "scientific researcher." Perhaps the word "scientist" can be used properly in the third person. As far as the term "student" is concerned, it probably cannot be replaced.

Dugin: *It is well known that there is such a concept as "student science." In your opinion, how well justified is the distinction between "adult" and "student" science?*

Gol'danskiĭ: In many areas of scientific searches it is simply impossible to work as an individual. Take, for example, modern experimental high-energy physics. The junior members of a group enter this field at the student level, but they essentially start by participating in the real science. If from the very beginning of the research the problem is described as small, that is, as a "student" problem, nothing useful and important will come out of it, because the problem has thus been restricted and the desire of the beginning researcher to study it has been repulsed. For the student it becomes simply another regular exam, to be passed and soon forgotten.

Dugin: *Vitaliĭ Iosifovich graduated from the chemistry department of Moscow State University. However, before this he studied at two other schools: Leningrad and Kazan' Universities. All three of these very old Russian universities affected the fate of the future Academician.*

At the ceremony when he was awarded the Aleksander P. Karpinskiĭ Prize established by the FVS Foundation of Alfred Toepfer (Hamburg, FRG) for distinguished Russian scientists in the building of the Academy of Sciences in Leningrad, Vitaliĭ Iosifovich said: "Leningrad for me is not just a great and beautiful city, it is a great part of my life, associated with unforgettable joys and sorrows, the happy time of childhood and elementary school years, my first friends and first student years, my first teachers and first loves; everything is familiar down to the smallest detail, and the houses and streets are so dear to me that it hurts. I remember the summer in the trenches in '41 and the winter during the siege. Here are the graves of my parents and the living memory of the heart. There is no other place on earth where it would be so thrilling to celebrate this solemn day as here, in my native Leningrad, on the banks of the Neva, inside the walls of the Academy of Sciences, two steps from the University."

In the prewar years, Leningrad University became for the student Gol'danskiĭ a holiday-like entry into the larger life, opening new horizons. Then the war came, and along with it grief at the loss of loved ones and friends, the construction of trenches and a bullet wound, searching for signal launchers for Luftwaffe bombers at nights, during the end of '41, work in the Komsomol district committee, hunger, the hospital, and then the Road of Life across the ice of Lake Ladoga.

He was evacuated to Kazan' and immediately resumed his studies at another famous university. This was for him a kind of resurrection after the rigors of the siege. Here he for the first time began combining work and study, spending the days working and the mornings and evenings in the university laboratory.

The USSR Academy of Sciences was evacuated to Kazan' during the war, so the students at Kazan' University had the rare possibility of attending the lectures of many famous scientists. The records contain the signatures of Academicians Aleksander N. Nesmeyanov, Pyotr A. Rehbinder, and other noted chemists and physicists. At the time he passed to the fifth course, Gol'danskiĭ was transferred together with the laboratory to the capital. While he was enrolled at Moscow State University, he became a junior scientific researcher of the USSR Academy of Sciences.

Again, as in Kazan', he combined attendance at seminars, lectures, and exams with intense daily work in the institute laboratory. This was essentially one of the first examples of combining science at the university and academy levels, and it ensured the successful defense of his diploma project in the summer of '44.

It is obvious that you have achieved great success in physics and chemistry. However, imagine for a moment that these sciences do not exist. What other field of knowledge would you have felt drawn to?

Gol'danskiĭ: From elementary school days I have loved history very much and wanted to become a historian. At the age of 12 I obtained a permit to use the Leningrad Public Library, and my favorite pastime was reading books on history, and especially old newspapers and magazines. I remember how interesting it was to take, for example, newspapers with different viewpoints from 1912–1917 and, by comparing their contents, trace the development of events at the threshold of the revolution. To this day I remember many historical dates and events. I am interested in everything related to this field.

Dugin: *If it is not a secret, did something happen in those years of childhood and youth which ultimately caused you to choose your profession?*

Gol'danskiĭ: Perhaps something in my thinking changed after I devoured with great interest the book *Chemistry and the Ways of It's Development [Khimiya i puti ee razvitiya]* by Boris Nikolaevich Menshutkin. I was most of all interested in the stories in it about the first discoverers of radioactivity. While still in the ninth grade, I thought of the question: how can the rays of radium on the one hand cure cancer, and on the other cause this most dangerous disease of modern times? I was captured by the idea that difference in effects is connected to different types of radiation. I wrote to the Academy of Sciences. They answered me. My letter had been given to a remarkable man, one of our greatest oncologists, Lev M. Shabad, later a member of the USSR Academy of Medical Sciences. He invited me to the laboratory, explained everything thoroughly, presented me a guinea pig, and advised me to study in the chemistry department. I finished school before I was 16. In order to go to college, I had to have special permission. Shabad again helped me here, and I became a student in the chemistry department of Leningrad University. I was not required to take the entrance exams—I had a certificate with a gold border; in those days medals for early graduates were still not established. Thus, one book and an unforgettable meeting became turning points in my life.

Dugin: *Apparently, you also met interesting people later?*

Gol'danskiĭ: Of course, there were many such meetings. I hope that sometime I will have the time to write about them. I managed to publish something in, for example, the *Reminiscences of Academician Lev A. Artsimovich.* Conversations with Landau always made a strong impression. If it had not been for the direct contact with the great German scientist Mössbauer, we would perhaps have chosen different methods for solving some of our problems and would have encountered great difficulties. Every meeting with the Nobel Prize winner Mössbauer provides a fresh stimulus for research in gamma-resonance spectroscopy.

As far as an image for emulation is concerned, greatness is, as is well known, most easily seen from a distance. One of the great people I would single out is Louis Pasteur. The book *Microbe Hunters* captivated me already in childhood. Almost every hero in this book of Paul de Kruif is attractive and vividly described. Among them is the great Pasteur, a lofty example for emulation, in spite of all his weaknesses. He was no stranger to conceit, for example. Sometimes he was very quarrelsome. And his arguments became very heated, sometimes even coming to blows. But along with this he was a great, brilliant personality. He saved the lives of people, and all his activity was motivated by the highest degree of humanism.

I know from my own experience that dealings with extraordinary people enriches one's understanding of the history of our country. For example, the spirit of the revolution became more vivid to me thanks to a single meeting with the legendary hero of the Civil War, Pavel Efimovich Dybenko (shortly before he was arrested and shot in 1938). Many years have passed since that meeting, but the impression remains just as clear as on the first day.

Dugin: *What qualities above all do you value in people?*

Gol'danskiĭ: Honesty and kindness.

Dugin: *What is alien to you?*

Gol'danskiĭ: Perfidy, perhaps.

Dugin: *Maybe you have noticed some qualities in yourself which you would like to get rid of?*

Gol'danskiĭ: Oh, yes, a great many of them. It is difficult to list them all.

Dugin: *What do you like about yourself?*

Gol'danskiĭ: My sense of humor (laughs).

Dugin: *What in life gives you particular satisfaction?*

Gol'danskiĭ: Perhaps interacting with people.

Dugin: *One often hears it said that a cultured person is one well versed in literature and art. There is also the broader concept of a cultured person as one who has reached the level of achievements of his time, the level of this century. What is your idea of a cultured person?*

Gol'danskiĭ: I think it is inaccurate and narrow to take the concept of "culture" to mean only literature and art. Culture must also include the sciences: philosophy and biology, physics and history, economics and medicine.... Acquaintance with the natural sciences was not considered necessary for a cultured person relatively recently. Evgeniĭ Bazarov, the Turgenev hero in *Fathers and Sons*, speaks in modern language, he is a technocrat, he consciously repels people of humanistic leanings. He limits his own range of interests to the natural sciences. Remember, for him nature is not a temple, but a workshop. The ideal modern person must strive to have broad interests. These days, when science is becoming so specialized, it is not simple to have a broad horizon. Nevertheless, in the ideal case (unattainable I'm afraid) a cultured person keeping up with the times must have at least some general idea of the fundamental scientific importance of research on how elementary particles behave at high energies, he must know to which areas of science pulsars and quasars, ribonucleic and desoxyribonucleic acids belong, and what gene synthesis is good for. He should at the same time recognize the difference between the Gothic and Empire styles in architecture, understand the language of the cinema, and be able to distinguish the works of, say, Antonioni and Kramer. A new exhibit, a book of poetry, a work of fiction—nothing should escape this highly cultured person.

It is clearly easier to try than to succeed in broadening one's horizon. The only way to do it is by constantly working on oneself. A feature common to every cultured person is engagement in daily, constant self-education, so that it becomes habitual. But of course it is important to avoid the danger of superficiality. These days, unfortunately, the breadth of a person's education often turns out to be measured in terms of a crossword puzzle rather than an encyclopedia. The gradual accumulation of knowledge in the end leads to a qualitative leap, when free association produces genuine taste, and familiarity with great works teaches one to distinguish genuine beauty from the false. Erudition generates understanding.

Dugin: *A joking question: What three subjects (no more allowed by the rules) would you take with you to an uninhabited island?*

Gol'danskiĭ: Such a question deserves a joking answer: I would take three friends of both sexes, who are energetic, skilled, and cheerful.

Dugin: *How do you feel after a great deal of stressful work?*

Gol'danskiĭ: With more than six decades behind me, I can see how this feeling has changed with the years. Earlier I felt joy and satisfaction. Now these are mixed with more and more sad notes.

Dugin: *Can you define the inner core which gives you the strength to endure the disappointments unavoidable in scientific work, as in any other activity?*

Gol'danskiĭ: I shall answer this extremely laconically: faith in human mind.

Dugin: *In attempting to penetrate the complex world of a personality, one nearly always risks seeing it with bias, one-sidedly. Reflecting upon this, I asked a person whose friendship with Gol'danskiĭ is so old it could be marked by a kind of jubilee to take a bystander's look at Vitaliĭ Iosifovich. I spoke recently with Nikolaĭ Markovich Émanuél', Hero of Socialist Labor, Lenin and State Prize winner, member of the Presidium of the USSR Academy of Sciences, and Academician-Secretary of the Division of General and Technical Chemistry of the USSR Academy of Sciences.*

> I have known Vitya, Vitaliĭ, Vitaliĭ Iosifovich for five decades, first as a laboratory assistant in the catalysis laboratory of the Institute of Chemical Physics, USSR Academy of Sciences, and then as a graduate student of Academician Nikolaĭ Nikolaevich Semenov.
>
> A variety of approaches to scientific creativity, a variety of experimental techniques and objects of study have provided unusual stimulus to forming the breadth of Gol'danskiĭ's views, his ability to instantly orient himself in new subjects.
>
> As scientific secretary of the institute, in 1947 it was my job to carry out the formalities confirming him as Candidate of Chemical Sciences after his brilliant dissertation defense. Whatever has happened to Vitaliĭ Iosifovich in the rest of his life can probably be heard only from him himself. This is a kaleidoscope of moving events,

problems, themes, a fireworks display of achievements and results. It is undeviating growth of mastery, scientific maturity and authority, and numerous indications of his talent and achievements at home and abroad.

I remember Gol'danskiĭ's defense of his doctoral dissertation on a nuclear topic in 1952. The Scientific Council was presided over by an eminent scientist who had brought fame to Soviet physics throughout the century: Academician Igor' Vasil'evich Kurchatov. The results obtained by Gol'danskiĭ were impressive. Later, working at the Lebedev Institute, Gol'danskiĭ carried out classical studies on the electromagnetic polarizability of hadrons and the determination of the proton polarizability constants, research on the Vavilov-Čerenkov radiation of cosmic particles in the atmosphere, prediction of the existence and properties of earlier unknown neutron-deficient isotopes, and the theoretical discovery of a new form of radioactive decay of nuclei: two-proton radioactivity.

In later years, having returned to the Institute of Chemical Physics, he laid the foundations and successfully developed a new area of science, nuclear chemistry, which studies the relations and effect of the chemical properties of matter and the characteristics of nuclear and elementary-particle transformations. The observed phenomena provide unique information on the structural and dynamical properties of the molecular and crystalline surroundings of these nuclei. He discovered the quantum low-temperature limit of the chemical reaction rate. The results he obtained are extremely important for solving the problems of chemical and prebiological evolution. In addition, Vitaliĭ Iosifovich is an excellent speaker and teacher, a university professor, and one of the directors of the All-Union Society "Knowledge" (Znanie).

I have not yet mentioned another side of Vitaliĭ Iosifovich's talents: his outstanding literary ability. He writes brilliantly: scientific articles, monographs, prose, verse, humorous sketches, winged words and phrases. For more than 10 years now he has published, under a pseudonym, his witticisms on page 16 of the *Literaturnaya gazeta* (Literary Newspaper).

Vitaliĭ Iosifovich is loved by scientists of all ranks, comrades, friends, and associates. In the fall of 1983 he shook us all up when he decided to end the exasperating problems caused by his stenocardia and boldly went in for heart surgery. Now he is again in good shape, working, meeting with colleagues, excited by new thoughts and ideas. I sometimes want to ask friends "Don"t push so hard for meetings with Vitaliĭ Iosifovich, let him get his health back," but it is impossible to hold anyone back. With Vitaliĭ Iosifovich everything is warm, merry, interesting, and very useful in terms of scientific activity. I would therefore like to conclude by wishing my dear friend Vitya Gol'danskiĭ—a man with a new heart—a new creative life, for at least another 60 years according to Mechnikov—to age 150, and à la modern gerontology even more. Especially since science is advancing swiftly and gerontology is not standing still.

Dugin: *What can be added to the opinion of a friend and colleague? The inquisitive mind of Gol'danskiĭ transforms everything around it into a topic of knowledge. Leningrad and Kazan', Moscow, the university, the labora-*

tory, meetings with famous scientists, celebrities, landscapes of Crimea—all this becomes a gigantic laboratory for him, in which the scientist seeks connections, compares, and analyzes. The gift of analyzing power is combined in him with the very rare ability to instantaneously switch from one topic to another.

In a photograph album at home, Vitaliĭ Iosifovich has a series of snapshots in which he and two of his friends were photographed in nearly identical poses, but at different times, over the course of many decades. "See how we have changed with the years," says Vitaliĭ Iosifovich, showing these photos. "Notice how this occurred not simultaneously, but how during a given period one changed more than the others."

And in this nearly insignificant incident he reveals the main want of an Academician: to study and compare. Even when he himself is both the observer and the object of observation.

The Art of Developing Science

The second half of the 20th century is noteworthy for the transformation from a manufacturing-technological society to a scientific-informational one. This is already seen in the case of countries like Japan and the United States. This means that the highest labor productivity will, in the end, be attained where it is possible to create the most efficient system of production and ensure the highest development of fundamental science.

In relation to this there is serious alarm at the state of higher education in our country and the importance of our achievements in fundamental science. Unfortunately, our lag behind the leading countries is quite obvious. And it is increasing. And our advantages, such as they are, are gradually being eroded.

A simple example is the awarding of the highest international prizes for achievements in the natural sciences, the Nobel Prizes. I shall not compare here the absolute numbers of such prizes obtained in postwar period by Soviet scientists (6) and American ones (more than 70). An important factor for this is that the Americans exert powerful pressure on the Nobel committee. I am worried about something else: of the six prizes awarded to Soviets, five have been awarded for prewar work, and the sixth for work done in the early 1950s. Of course, we have produced works at the Nobel level since them, but the amount of such works in the 1960s through 1980s has been much greater abroad (in the United States, Britain, Germany, France, and so on).

The situation with science here in the 1960s through 1980s reflects the stagnation in our society. For example, have any large-scale fundamental scientific studies been financed? Priorities have often been determined not by the genuine value of some gigantic technical installation or by some area of research, but rather by the desire to have short-term, record-setting (more precisely, pseudo-record-setting) payoffs. Science has been used for propaganda purposes, to demonstrate an alleged superiority over the West. This has happened not only where such superiority actually existed, but also where it never did in the eyes of serious scientists around the world. This has led to areas closed to criticism and objective evaluation, and fruitless expenditures of enormous material and our best human resources.

We have lived for long years according to the principle of self-training: "Close your eyes and keep repeating: 'I am fine, I am warm, and you will be fine and warm'." Only in the last three years have we finally decided to open our eyes wide and soberly regard the situation in our country and the world. But still now many dogmaticists cannot and do not want to break the habit of keeping their eyes closed.

Long ago, Academician A. N. Nesmeyanov, president of the USSR Academy of Sciences, compared the development of science to the battle for possession of a building. The first stage is breakthrough to a new floor, and the second is dispersal throughout the floor. The breakthrough to a new floor is the main task and activity of fundamental science. As a rule, such breakthroughs are unpredictable, but it is they which bring in their wake genuinely revolutionary achievements not only in the area of the breakthrough itself, but also in many related sciences, in their practical applications, in the scientific worldview itself. The work on inculcation of the results of new discoveries, on their technical mastery, are typical examples of "dispersal throughout the floor."

The largest fraction of our scientific-organizational activity in the postwar years (including such remarkable achievements as the harnessing of nuclear power and the exploration of outer space) has been concentrated on problems of "dispersal throughout the floor." "Breakthroughs to a new floor" have gradually ceased altogether to be viewed by the USSR Academy of Sciences as its main area of endeavor (perhaps because such breakthroughs cannot be planned!). At assemblies of the academy one almost never hears anything about scientific discoveries [except for so-called "patent" discoveries, of which we (like no other country in the world) have hundreds officially registered by the State Committee of Inventions and Discoveries, but these do not in the slightest determine international scientific progress].

The USSR Academy of Sciences has not been able to ensure the rapid growth of fundamental research; great confusion has developed in defining and distributing the roles and tasks of the academy and department ministries in scientific-technological progress. In the end it has become completely unclear who is primarily responsible for the breakthroughs.

In speaking of the state of development of fundamental science, we are focusing on the natural sciences. However, the same can be said even more strongly of the miserable state of our social sciences. Administrative-fiat methods of direction and control together with dogmatism have caused great damage in the social sciences. History, economics, sociology, psychology, philosophy, law—the existence of all these fields can be spoken of only with great reserve. One of the main problems of philosophy, as is well known, is how to comprehend the achievements of the natural sciences, how to bring them down to the level of social categories. We have been dominated by a natural philosophy whose main goal was abuse and denial of everything new in the natural sciences and attempts to impose upon nature the falsely understood positions of dialectical materialism (similar to the way that the falsely understood, vulgarized tenets of historical materialism have been imposed on society). As a result, in the post–October Revolution period the scientist-humanitarians of the capitalist countries made much greater achievements in comprehending the meaning of the laws of nature and the origin of humanity itself. Meanwhile our scientists—natural scientists and humanitarians—in many respects completely stopped understanding each other (recall the attempts to annihilate genetics, cybernetics, quantum mechanics, relativity theory, and so on).

The war between conservatives and innovators, dogmatic and creative principles, old and new in science (perhaps particularly in the social sciences) is no less acute than in the creative unions. After all, inertia and overcautiousness, fear of losing clearly undeserved, exclusive, former privileges are manifested openly fairly often also in science.

Whereas in the past the USSR Academy of Sciences was generally recognized as the institution which was the most democratic and independent of the official apparatus, and won valid approval for this, now, contrasted with the background of ever-growing glasnost, the academy more and more often resembles a retrograde, official institution, reminiscent of some ministry of applied science.

After all, it was the academy in the 1960s that dealt the first blow to Lysenkoism (which cost Aleksander N. Nesmeyanov his position as president), and in the 1970s actively resisted pressure from above on elections, corruption, and the automatic selection of high-ranking "officials" (among the candidates were the head of the Science Division of the Central Committee of the CPSU, government ministers, lawyers from the Vyshinskiĭ "school," philosophers hardly worthy of the title, and so on). The importance attached to opinions of members of the academy at that time was determined primarily by their scientific accomplishments and real merits, not by their rank or post.

Now the situation is different.

At the head of the academy has arisen a real, multilevel hierarchy, hyperinflated in numbers and supported at present by a unique system of "differentiated democratization" which has three levels.

The highest echelon of the academy is the president, 10 (now temporarily nine) vice presidents, the academicians-secretaries of the 19 divisions according to specialty, and the members of the presidium: altogether about 50 people. The age limit here is 75 years, but it has been in effect only since 1990, and only for future members of the presidium.

The elections to this highest echelon, once every five years, are held without competition, with one candidate per vacancy, and "ordinary" academics essentially do not participate in nominating candidates; they usually learn the names of these candidates only at an assembly of the division or the academy.

Therefore, a senseless and humiliating situation has been created which completely contradicts the present-day orientation toward democratization, glasnost, a pluralism of opinions, and this situation in many respects transforms the elections for directorship of the academy into a meaningless function. Just think: these are the elections of the most important organizers and leaders of our science among the only 300 scientists who, according to the statute of the academy, have enriched science by works of primary scientific importance. Why is it that among these "sages" there is exactly one—no more, no less—candidate worthy of performing in each field the daily tasks of running the Academy itself, why is it that voter-Academicians "from the threshold" are not allowed to judge who among them can most successfully carry out such tasks, while this judgment and these decisions are *a priori* assumed to be the privilege and monopoly of the high authority; why must the ruling council of the academy be chosen by someone higher, not as alternatives presented for judgment and selection by Academicians, but as a single, *a priori* specified candidate?

Many members of the academy remember how three years ago in the elections to the presidium of the academy the proposition of only a single alternative candidate among 50 candidates was enough to get the person making the proposition, a very worthy scientist, sharply rebuked by the Secretary of the Central Committee of CPSU sitting in the presidium of the assembly.

The leadership of the academy has in the last couple of years taken a number of clearly useful measures to expand the rights of the divisions, with the aim of endowing them with the functions of full-fledged managers of the institutes making them up.

Unfortunately, this has not occurred without new bureaucratic inventions. The presidium apparatus of the academy has acquired new bureaucrats, the unelected posts of "aide" to the president and vice presidents, which has taken the academy another step farther along in the direction of a ministry, and raised yet another barrier between the scientists who run the academy and the scientists who are members of the academy.

The senior managerial staff are institute directors (several hundred people). The age limit for this position is 70 years for academy members and 65 for others. But this limit takes effect again from the time of discharge of the plenipotentiaries elected previously. Last year, after many articles in the press, a general assembly of the academy decided to elect the directors democratically with the participation of the staff of the institutes. However, shortly afterwards the presidium of the academy, without prior permission, abolished this competitive system for 10 institutes in which the new directors were designated by fiat (which, incidentally, contradicts the statutes according to which institute directors are elected by the academy divisions). The total term of the director plenipotentiaries, the number of allowed five-year "terms" of their tenure, is not restricted.

Here and there a limit on the total term of the plenipotentiaries has been introduced, but, however strange it may seem, this has occurred only for those posts which are unconnected with any sort of administrative power: for example, for the editors-in-chief of scientific journals, for chairmen of scientific councils (who have only consultative, advisory functions), and so on.

On the threshold of the 19th All-Union Party Conference there was wide discussion in the press of measures needed for the further democratization of our state and party life. It is impossible not to think about the much more important measures of this type in science. I see the following as some of them:

- The introduction of compulsory elections on a competitive basis (with several candidates) and substitution (no more than two five-year terms) for tenures in the highest governing posts (president, vice presidents, academicians-secretaries, presidium members) and in the post of director of institutes in the academy and republican academies.
- It would be good to start a newspaper—e.g. *Soviet Science*—which could become an effective means of explaining the role of science in the spiritual development of our society, and act as a forum for expressing public opinion on important questions in the life of our science, its vital—thematic and organizational—problems. Such a newspaper could introduce discussion of the role of Soviet science in perestroika, in the democratization of our society, on the ecological aspects of science, on questions of scientific ethics, and so on. *Soviet Science* could also publish materials devoted to the activity of Soviet and foreign scientists in the struggle for peace, against the nuclear threat, and articles on history, including the history of science. The newspaper would become a great aid in attracting scientists to active participation in perestroika, in its scientific and technical justification. It is difficult to exaggerate the possible value of a newspaper of Soviet scientists in these days, when the mass media have taken on the role of the locomotive of perestroika train, pulling this train to our common goal and also raising the alarm when old dogmas stand in the way and are slow to move aside.

Struggles of a Soviet Scientist[a]

The extraordinarily frank observations of Vitali I. Gol'danskiĭ, a prominent member of the Soviet Academy of Sciences, were first published in October, 1987, in the Soviet weekly *Ogonyok*. Several months later, when he was visiting the United States, Gol'danskiĭ passed a copy of the interview to Paul Doty, director emeritus of Harvard University's Center for Science and International Affairs. Doty forwarded it to the *Bulletin*, describing it as "a very revealing disclosure of how glasnost interacts with a very important and courageous Soviet scientist."

Gol'danskiĭ, *Ogonyok* says, "was one of the first to speak about the deep connections between scientific-technological progress and new thinking." The Lenin Prize winner, a leading chemical physicist, is the new chairman of the Soviet Pugwash Committee and is a member of the Committee of Soviet Scientists for Peace and against the Nuclear Threat. The interview, which is excerpted here by permission of *Ogonyok*, was conducted by Vanda Beletskaya and translated by Kurt Campbell of the Harvard center.

—The editors,
The Bulletin of the Atomic Scientists

Gol'danskiĭ: The struggle for a non-nuclear earth is the common business of all inhabitants of the planet; nuclear war is their common death.

Thus scientific-technological progress is becoming a prerequisite for the formation of new thinking in international politics. The Soviet leadership, specifically Mikhail Sergeyevich Gorbachev, was the first to realize the necessity of such new thinking. "Nuclear war would be a catastrophe for everyone without exception," he stressed. From this flows the deep scientific basis of Soviet foreign policy and our program of "disarmament for development" as opposed to "armament instead of development."...

Beletskaya: *Is an accidental, unsanctioned, unintentional nuclear explo-*

[a] Interview by Vanda Beletskaya.

sion possible? Let's say, in a missile accident or that of a separate nuclear warhead?

Gol'danskiĭ: One of the tasks that confront nuclear weapons designers is precisely the reduction of such a possibility to a minimum—the creation of a mechanism which could not explode accidentally or unintentionally. But all the same, that the probability of an accidental explosion is equal to zero, not to some very small but definite number but precisely zero—this I cannot say. Especially after Chernobyl. Before Chernobyl the question would have been answered differently than today. We must extract a lesson from what has happened. This also is included in the concept of "new thinking." We must not, we do not have the right to blindly or boundlessly trust technology no matter how perfect it becomes. And we must not forget the colossal responsibility which rests on the shoulders of those who have such technology at their disposal. Technology should be as foolproof, as idiot-resistant as possible; and people should be as conscientious as possible in their handling of it... .

Beletskaya: *We constantly speak of the necessity of new thinking in the nation's foreign policy; but what about domestic policy?*

Gol'danskiĭ:The two central directions of scientific-technological progress—information processing and the breakthrough into space—have become very important elements of contemporary productive forces. The bringing of social relations into line with productive forces in our time inescapably includes such important aspects of new thinking as democratization and glasnost.

Currently the flow of information is growing extraordinarily quickly. The quality and means of its registration, storage, and transmission are being perfected at the same rate. Computer languages—Algol, Fortran, Pascal, and others—are becoming a sort of Esperanto of our time.

Undoubted realities of the near future are planet-wide radio and television programs, global communications in machine languages, the reception of information by individual computers and its processing by various types of programs.

Beletskaya: *That seems almost fantastic! But what about our scholars, in this country, are they prepared for this?*

Gol'danskiĭ: Alas, they are far from where they should be. And the reasons for this are not purely technical but, if you wish, also ethical. They are found in the long absence of such concepts as democratization and glasnost.

For example, scientific journals are a vitally necessary source of information for scientists. But it is becoming more and more difficult to gather information from them about the work of our foreign colleagues. Subscriptions to foreign journals are being reduced due to difficulties with foreign currency, but that is not the issue.

We have huge problems with our photocopying. We have few photocopying machines and the rules for their use are extremely complicated. Their quality is often very low due to their dilapidated condition and shortage of the imported powder which is necessary for their functioning. And this is just the beginning! After all, in order to know precisely what article in what journal you need to read or copy, it is necessary first to see the journal itself or at least its table of contents. We receive patently too few journals from abroad, and what's more, they are concentrated as a rule in Moscow and Leningrad and to a lesser degree in the regional branches of the Academy of Sciences. At almost all our institutions of higher learning there is the strongest hunger for foreign scientific journals. VINITI [the All-Union Institute of Scientific and Technological Information of the Academy of Sciences of the U.S.S.R. and the State Committee of Science and Technology of the U.S.S.R.] could help by distributing photocopies, if even of the table of contents alone, from published journals. And it is doing this (for which I thank it), even if only for a very small segment of our libraries and scholars, in particular for the members of the Academy of Sciences personally. In this regard, however, unknown bureaucrats painstakingly keep us frequently from professionally important information which we urgently need and which, of course, is completely accessible to our foreign colleagues. Journals like the very important universal sources of express information about all the natural sciences like *Nature* (Britain), *Science* (United States), and others arrive with missing pages. And from the photocopies of the table of contents of these journals, important sections—such as "The Latest News," "News and Commentary," and so forth—are frequently entirely or almost entirely removed.

Beletskaya: *I wonder precisely what these bureaucrats manage to transform from science into secrets to be kept from Soviet scientists?*

Gol'danskiĭ: I checked up on that in foreign libraries. Here are a few examples of "illicit" titles in the issues of *Science* from the 8th and 15th of May, 1987 which were struck from the table of contents: "Glasnost Comes to Soviet Physics," "Nuclear Power after Chernobyl," "Hungary Seeks Ways to Live off Its Wits." And here is the title of a book which was removed from the list of book reviews in *Science* from the 15th of May—*The Communist Party and Soviet Science.*

Beletskaya: *As they say, it speaks for itself.*

Gol'danskiĭ: Precisely. And there are dozens of such examples: articles (and even the simplest titles of articles) about seismic monitoring of nuclear testing, about various aspects of the notorious SDI, articles which contain in their title the word "Chernobyl." ... I looked at these "criminal" materials. They contain, in particular, a solidly scientific refutation by American seismologists of insinuations to the effect that the size of certain of our underground nuclear explosions (which were conducted before the recent moratorium) exceeded the established threshold of 150 kilotons; a refutation of frantic rumors about the high levels of radioactivity in Western Europe after

Chernobyl; and skeptical judgments by prominent American scholars about the project on the pumping of x-ray lasers from nuclear explosions.

Beletskaya: *Now why in the world is it forbidden for Soviet scholars to know about all of this?*

Gol'danskiĭ: Nobody knows. Probably for the same reason that in the Natural Sciences Library of the U.S.S.R. Academy of Sciences, issues of *Nature* which even once refer to Chernobyl were not available for readers. And as a result, our scholars, when participating in international meetings on a range of actual problems, were much more poorly prepared for the discussions than their Western colleagues and also than scholars from other socialist countries. (After all, such removal of material from academic journals does not occur over there.)

That's the way that the past—with its decrepit, nonsensical prohibitions, even while verbally acknowledging all of the slogans of reconstruction and renewal—erects barriers, in effect impeding scientific-technological progress, depriving Soviet scientists of necessary information which is, by the way, completely accessible not only to foreign scholars but even more so to intelligence officers.

After all, the means for monitoring from space everything that happens on the surface of the planet (and at times underwater or beneath the earth's surface) with the aid of spy satellites equipped with every possible type of sensors are currently being developed extremely rapidly. For example, recently *Scientific American* published diagrams of how precisely the shape of a tank can be worked out in detail in recent years from heights of several hundred kilometers.

All the nations, all the peoples of our planet have become to a certain extent like adjoining rooms in a communal apartment and their inhabitants.

Whether this is good or bad is a special question. Whether we should praise or curse science for all of this is not the issue here. This is all given, an integral feature of the times in which we live. It is a reality with which we must deal, by which we must be guided so that we do not wind up hopelessly behind the times not only in technology but in our way of thinking... .

Beletskaya: *I am under the impression, Vitaliĭ Iosifovich, that in your childhood and youth you studied history and were even unsure of what profession to choose. What is your opinion of the current condition of our historical sciences?*

Gol'danskiĭ: In my opinion the historical sciences still lag behind literature and art in the depiction of the heroic and tragic pages of the life of our country and party. We are accustomed to history being reshaped and altered time and again to suit immediate interests.

Fifty years ago the procedure for such reshaping was simple: at the regular lesson in school we were told which portraits to cut from the textbook today and which names to cover over with ink.

Thereafter these procedures grew more complicated and effective, but their essence remained unchanged. As a result, history today frequently reminds one of a canvas onto which new images have been repeatedly painted on top of the previous ones. This has created the difficult and pressing problem of restoring the original pictures of history. Such a restoration cannot be accomplished without primary sources. Without them history cannot be seriously studied nor taught: it is impossible to understand today without first understanding yesterday.

You are correct. From childhood on I very much enjoyed history and wanted to devote myself to that science. Having learned of this, an American colleague about 10 years ago gave me as a gift a huge, thick book entitled *Page One*—facsimiles of the most important front pages of the *New York Times* from 1920 to 1975. How interesting and instructive such a book is! How vividly it shows, for example, just how little attention was devoted in the U.S.A. to the decisive contribution of Soviet Armed Forces to the crushing defeat of Hitler's Germany! But where among us is there an analogous book? Before 1956 even John Reed's historic work *Ten Days That Shook the World* was strictly forbidden. In our encyclopedias it's impossible to find dozens of names of people who played a significant role (granted that it might be a predominantly negative one, but nevertheless they lived and acted and their names frequently figure in Lenin's works) in the history of the party and nation. Two of the 10 leaders of our government during the 70 years of Soviet power, for example, are absent—and there was a time when that went for three or even four of them. The common reader does not have access to the files of old Soviet newspapers and journals which are kept in a small number of libraries. He does not have access to the living pages of the history of the revolution, the civil war, the 1920s and 1930s, the Great Patriotic War, or the postwar years. Many old newspapers and journals have become, it is true, dilapidated—that is precisely why publications of the type of *Page One* would prove to be useful. Or stenographic accounts of party congresses, which you will also never find no matter how long you search.

We simply do not comprehend how large an educational role such living pages of history could play for our young people! In my opinion precisely now it is urgent to publish the report which [Nikita] S. Khrushchev gave at the Twentieth Party Congress on February 25, 1956. After all, that document has become so memorable in the life of our generation. On the other hand,

[1]This refers to a staged meeting of the All-Union Academy of Agricultural Sciences which provided a forum in 1948 for the influential biologist and agronomist Trofim Lysenko, whose discredited theories severely set back Soviet agriculture and science.—eds.

the republication of the stenographic account of the disgraceful August session of the VASKhNIL[1]—that "benefit performance" for the academician Lysenko—or the monstrous "judicial speeches" of the future academician Vyshinski[2] would be an instructive and caution-inspiring example for many, including scholars.

Let's examine recent events. Certain "sages" are turning up who look at all of our calamities—Chernobyl, the [sinking of the steamer] *Admiral Nakhimov*, the mine explosions—and see in them proof that if the nation has changed, then it has changed for the worse. Calamities have not increased. Rather, glasnost, honest and frank information about these calamities, has. If glasnost had existed even earlier, then there would be less laxity, fewer catastrophes and human deaths.

The practice of keeping silent about the truth, of concealing or distorting undesirable information—which existed before [Gorbachev became general secretary in] March 1985—not only greatly harmed science and the economy, not only promoted moral erosion, but was also a manifestation of a libertarian antiscientific approach. It was in direct contradiction to the requirements of scientific-technological progress and the rapid and explosive development in such important directions as information processing, radioelectronics, and the exploration and utilization of space.

An active builder of the future cannot help but sense the deep kinship, the indissoluble harmony of his interests with the interests of the Motherland—with its prosperity and well-being, whether concerning the economy or the conservation of nature, history, or cultural monuments. At the same time he must be able to think independently so as not to take instructions from higher comrades automatically for an expression of the true interests of the Motherland. In this regard it is extremely important that Soviet people be fully informed, that they be provided with the fullest—with the single exceptions of pornography and misanthropy—volume of information.

---

[2]The prosecuting speeches of Andrei Vyshinski in the Great Purge trials of 1936–1938. —eds.

Why Are We Lagging Behind?

Remarks about Fundamental Science[a]

The phenomena which characterized the period of extensive development of the productive forces in the country very strongly affected the development of science and technology and significantly retarded the rate of scientific-technological progress in our society. As a result, not only did we fall behind in several fundamental areas of technology of great importance to the national economy and national defense (microelectronics and information technology, mechanical engineering, biotechnology, medical technology, and so on), but we allowed a serious lag to arise in the development of fundamental science itself. This has happened in spite of huge, ever-increasing material expenditures to the network of scientific institutions.

For example, we have constructed the largest optical and radio telescopes in the world, but have not made a single first-class astronomical discovery with them. In spite of our powerful particle accelerators, we have not become the leader in a single area of elementary-particle physics. Unfortunately, this list could be continued. What should be done? Why is our contribution to world science now not proportional to the strength of our government and the talent of our people? We must find an honest and unbiased answer to this question.

The unpredictability of the specific content of important discoveries in fundamental science does not at all imply that it is impossible in principle to predict the general trends in its development; such predictions can and should be made.

[a] With Yuri Osip'yan.

In the United States, as a rule, the trends in the development of science are analyzed by many special centers and organizations. From time to time we also carry out organized campaigns to make such predictions, but our approach to them is insufficiently systematic.

We essentially lack serious comparison of predictions and reality, analysis of which predictions were correct and also why others were incorrect, critical use of experience in comparing predictions in the past for further extrapolation. Our academic divisions, scientific councils, and institutes are not required to do this, and as a result the prognostication is not specific enough, and the prognoses are made on the basis of figures which are poorly justified and inconsistent with financial planning agencies.

To put it briefly, serious improvement is needed in the prediction and long-term planning of science. It is necessary to drastically reduce the quantity and change the character of the materials required from academic divisions, scientific councils, and institutes, and to pay serious attention to the analysis of earlier predictions and their comparison with reality.

The second group of questions whose solution may be no less important for accelerating scientific-technological progress in our society pertains to the processes by which scientific and technical developments enter into production. This is the most important factor determining the level of present-day technology and is of critical importance for our national economy, but it is very inefficient and requires the overcoming of enormous obstacles and numerous contradictions. The question again arises: why? Why in the West is it possible for manufacturers to get new scientific developments directly from the hands of scientists, while here they have to be "inculcated," by importuning ministries and departments?

After analyzing these problems, one arrives at the conclusion that the answers to the two groups of questions about the lag in fundamental science and "inculcation" are the same. The roots lie in the incorrect, inefficient system of relations between science and technology, enhanced both by decisions of the state and by the general ideology in our country, which is obstinately cultivated by a significant fraction of the governmental and industrial complex and, seemingly paradoxically, by many leaders of our highest scientific-technological institutions.

Let us explain what we mean. In spite of the fact that the ultimate global goals of science and technology (of production) are universal—to serve humankind and social progress—the specific goals within the overall general process differ.

The problem of science is to learn the laws of nature and society, to obtain objective, reliable knowledge.

The problem of technology is the efficient use of the knowledge obtained by science for specific goals in the development of production, social progress, and, ultimately, the universal good of humanity.

Accordingly, the agencies controlling science (the USSR Academy of Sciences) and technology (ministries and departments) must have different responsibilities to the state.

The Academy of Sciences is fully responsible for the status and level of science in the country and for scientific training. However, it cannot be directly responsible for the level of development of technology and production in whatever branch of the national economy of the country, releasing the branch itself fully or partially from this responsibility. At present these boundaries are artificially mixed up.

It is considered correct that science and the academy, in particular, along with ministries and institutions, be required to carry out specific exploitations and to introduce scientific results into the national economy. This is precisely how the top level decisions about specific questions of the development of certain directions in the national economy are made. To what has this led?

The directors of the academy and its institutes are losing the feeling of responsibility for the development and level of fundamental science itself in the country. It is now a long time since we have been asked and forced to answer why we have allowed such a serious lag to develop in fundamental science. In return, it has become possible for us, together with the branches, to be evaluated on the basis of how well specific scientific-technological exploitations are realized.

On the other hand, the branches, with the blessing and aid of the central agencies, are trying to share with the Academy of Sciences the responsibility for the specific development of individual technical problems. This actually leads to demobilization of the scientific-technological potential of the branches. This is clearly seen in the example of the development of computer technology in this country, and, in particular, the development of personal computers. It is no secret that in this area we are far behind not only the developed capitalist countries, but also countries like Norway, Thailand, Taiwan, Singapore, and South Korea. Who should take responsibility for this and correct the situation? Of course, the corresponding area of industry, and no one else. But it points to science. And the academy needs to create a special primary institute to solve this purely technical problem and to involve several other of its institutes in the matter. Thus, the principal performer is found: the USSR Academy of Sciences, and the branch itself only helps it.

In Norway, Thailand, and Singapore there are no scientific institutes devoted to this problem, but still millions of personal computers are produced, and for the time being we must buy ours there. This is a typical example. A technical problem is elevated to the level of a scientific one. The Academy of Sciences forms whole divisions for this area, elects new Academicians, not because they have solved the problem, but because they are supposed to deal with its solution (i.e., there are indirectly stimulated just those who have ruined the problem). But this is not the main thing. The main thing is that the academy is distracted from working on fundamental problems, which has negative long-term consequences.

The branches do not develop the qualification and possibilities of their own employees, they do not present them with serious and important problems to solve, which in addition leads to a decline in the prestige of engineering work. We are on the threshold of another similar problem: that of magnetic information storage for computers. The annual sale of magnetic storage elements in the United States now amounts to 40 billion dollars. This is greater than the sale value of all semiconductor microelectronics elements. We do not have a single institute in our industry which could be involved in this area. It is not impossible that the USSR Academy of Sciences will be charged with this new problem and become responsible for solving it.

Therefore, in our opinion, the only route which can lead to an increase of efficiency in both science and technology is the separation of the responsibilities for science (the USSR Academy of Sciences) and for technology (the branches). This is also the way to put the economic mechanism into effect.

The branches fully responsible for technological problems will need to turn to science to solve unusual problems, and will devote some of their resources to science. Only then will natural and correct relations between science and technology be created for the solution of applied problems, and the problem of "inculcation" will disappear.

Today technology knows that science is obliged to deal in this way with technical matters. There is no need to attract science, it will itself beg to use its results. This is what leads to the ugly problems of inculcation. If science is freed from this responsibility, technology may in some cases not want to involve science in solving its problems, but may want to do everything by itself. Then there will be a single task left for many scientific institutions: occupying themselves with fundamental research. Then it will immediately become clear who is capable of doing this well, and who is not.

Of course, we do not want science and the institutes devoted to it to become alienated from working on technical problems whose solution is needed by our society today, nor do we want scientists to acquire an "easy life." On the contrary, the problem is to include science in economic processes most efficiently. But here it is necessary to find an optimal mechanism that works naturally. We think that our proposed solution of this very important problem best corresponds to the spirit of perestroika.

Science Has Had Its Say

I remember how we argued 30 years ago. Landau, pensive and even somewhat melancholy, was silent. And only after all the arguments of both sides had been "laid on the table" did he sum it all up: "How happy I would have been if the Snow Man had really existed—we do not have enough miracles in life! But, alas, he doesn't."

The same can be said today of the famous shroud of Turin. How gladdened I was in 1978 by the flood of reports in the most authoritative scientific journals about the surprising, unexplained features of this relic, with what interest did I read several years later the book by John Heller *Report on the Shroud of Turin.* Knowing of my interest in the shroud, a friend of mine from Brookhaven National Laboratory, Garmon Harbottle, regularly wrote to me about the preparations for the experiments on the radioactive-carbon dating of its age and on the difficulties of obtaining all the necessary permissions from church officials.

Finally, the disappointing, unambiguously negative result: the cloth the shroud was made from dates from the 13th century. What a great pity, we do not have enough miracles in life!

But I cannot accept the "explanation" of Lev L. Regel'son at all. He himself feels that it is completely unsound physically. In fact, there remain many unanswered questions about the amazing mastery of the unknown artist of the 13th century who created the shroud. Answers to these questions should be sought using the full arsenal of 20th-century methods, the laws of physics and chemistry, the experience accumulated by these sciences. However, I suspect that since the date the shroud was created has been moved up closer to the present by thirteen centuries, it will not be easy to find art historians and criminologists who will want to sort out the history of its creation.

Let Common Sense Triumph

There is a certain magic to numbers, and for some reason the year 2000 fascinates us. I should tell you that I love history and became a chemical physicist probably by accident. Books about the Middle Ages in Europe tell of how the year 1000 was awaited with terror: it was thought that it would coincide with the end of the world.

However, at that time humankind did not have the power to bring about the end of the world. Today the development of civilization has reached a level such that people do have the power to completely annihilate civilization. For the first time ever in the existence of humankind it is possible to carry out an evil "end of the world" project and to physically exterminate both humankind and civilization, and the planet on which they exist. This is the enormous difference between today's millenium and the earlier one. In this sense the magic of numbers works, especially now that people have become so susceptible to any sort of miracle, they can be made to believe in anything.

The variant with the end of the world is a definite reality today, but, fortunately, with the beginning of the 1990s, the last decade of our century, it is becoming much less probable than five years ago. I see two alternative paths of development of Europe and the world in the 1990s. If the ideals of perestroika, the new political thinking, the ideals of an existence in which common human values are the priority, are realized, this will in the end lead to the creation of a common European home and eventually to the creation of our common planetary home.

But let us not look too far into the future and instead consider the creation of a common European home in the 1990s. All Europe is now undergoing a renaissance. Its civilization is a completely unique compound. There are here individual diamonds—Russia, France, Germany, Britain, Italy, and the smaller countries. It is also a community. The great civilizations of Europe, unique in their history, politics, and culture, have crossed paths many times in the past and will continue to do so in the future.

The common European home is a building with many apartments, and the attachment of each inhabitant to his own apartment will not interfere with his attachment to the entire building. Russia is one of the apartments of this building most rich in culture. Take our literature: Russia can be considered the home of the library of the common European building, just as Germany can be considered the home of music and Italy the home of painting.

It must also be born in mind that even though each apartment of the common European building will keep its own rules and traditions, all will be in accord with the universal declaration of human rights. Their material levels and food-producing abilities will equalize and will be adequate for the entire building. All the inhabitants of the common European building, and even of all the world, will have to live in peace with nature. In the end, to live in peace with nature means living in peace with one's self. I hope that we will move in this direction, and I would very much like for our country, by the end of the decade, to have enriched the very bright spectrum of the future cooperation of the European peoples, or at least that it will not be a parasite of them. This is the first alternative for the end of the millenium.

To imagine the second variant of the end of the millenium, it is necessary only to rename Orwell's famous book *1999* instead of *1984* and to add something which the author only vaguely predicted: ecological catastrophe. In the second, Orwellian-variant a polluted society will certainly also imply a polluted nature. It is impossible to imagine such a world in harmony with nature. This alternative will occur if those full of hatred, envy, and greed—the most destructive human qualities—gain the upper hand.

Which of the two variants is realized in Europe will depend to a very large extent on the future of our country. If we remove the blinders from our eyes, if we reject living according to dogma, if we return to moral principles, if the freedom of the word and the freedom of conscience triumph in our country, if our economic structures are ruled by common sense, then will not only Russia be transformed, but so also will all of Europe be. The fate of the world is inseparable from the fate of our country.

Soviet Scientists Face Perestroika's Challenge

[Editor's note, *The Scientist*: When the democratically elected Congress of People's Deputies met for its second plenary session in Moscow in December 1989, delegates expressed frustration with the slow pace of reform. Among these frustrated delegates were scientists and academics, many of whom had been chosen by their colleagues in special elections at their institutions. Under the leadership of Vitaliĭ Gol'danskiĭ, director of the Institute of Chemical Physics in Moscow, they formed an Academical Deputies group and issued a public declaration of their aims. Gol'danskiĭ, a nuclear chemist, was awarded the Lenin Prize in 1980 and was named to the prestigious post of Academician in 1981. He has recently done much of his work in tunneling phenomena. Russian scientists have a long history of involvement in politics, a tradition that continues today as great political changes roil their lives and their careers. Following explanatory comments by Gol'danskiĭ, the declaration is excerpted below.]

Among the most important events in the political life of the USSR in 1989 were free elections to a new Soviet parliament, called the Congress of People's Deputies. For the good of the country, this new body must be an effective supreme organ of government, and not the rubber stamp that the Supreme Soviet was for half a century.

In the Soviet parliament, as in the entire Soviet society, one can now find a very wide range of political views—from Stalinist, ultraconservative, even Nazi-monarchist antagonists of Gorbachev's perestroika, to extremely left-wing radicals, who are becoming increasingly skeptical toward Gorbachev because of his compromises and concessions to the right-wingers.

As a major prerequisite for social and political consolidation, it was extraordinarily important to organize the scientists and scholars elected as people's deputies and to develop a general platform. In compliance with this main task, in 1989 an Academic Deputy Group was created. This group unites more than 100 deputies, including deputies from the Academy of Sciences of the USSR, academies of the individual republics and the Academies of Medical, Pedagogical, and Agricultural Sciences.

Having reviewed the country's current situation and its prospects, scientists, and scholars presented their point of view in a public declaration, which is introduced below.

It is important to stress what unites the absolute majority of scientists and scholars who are people's deputies: We are total antagonists of the right-wing opposition to Gorbachev, which we think not only endangers perestroika, but also makes very possible a return of the Cold War. Even when we are dissatisfied with Gorbachev's deeds and consider them inconsequential, irresolute, and even playing into the right-wingers' hands, we believe it is advisable to refrain from attacking him and thus open a second anti-Gorbachev front. We do not have another alternative, and it seems to us that the victory of Gorbachev's "new thinking" is the last chance in the 20th century to see our homeland become an equal member in the community of civilized countries.

We hope that the world scientific community will give us full understanding and support of our platform:

1. The country is going through a crisis stage in its development. A crisis is always a severe trial. But it is not a catastrophe or death as long as the society clearly understands the genuine causes of these difficulties, and sees their solutions.

We have certainly made progress.... Democratization is gathering strength, and a tendency to treat economics reasonably and effectively is increasing. A spiritual release from run-down dogma and a return to genuine moral values are becoming apparent.

There is a revolution under way in this country. We know that a revolution is worth something only when it knows how to defend itself. Today we must defend our perestroika and consolidate all democratic forces of the society around it. Defend it from whom? First of all, from all extremists, who are pushing our society into a quagmire of chaos, animosity, and violence. We believe that an increase of the right-wing extremism and attempts to bring us back into preperestroika times... are becoming more evident. The particular danger for the very existence of our country is the fact that the alliance between certain groups, who are guided by pseudomarxist stock phrases of yesterday, and ultrarightists, who borrow their ideology from even more distant times, is being strengthened.

2. What is, in our opinion, the medium nourishing modern extremism, the main obstacle to consolidation of all the healthy forces of society and the most important reason for the aggravation of ethnic problems?

We have failed to stop the deterioration of our economy. The command-economy has become bankrupt, but nobody has created another system. This is not because nobody knows how to create it, but because of the passivity, the selfishness, and the sabotage of the bureaucracy, and the ignorance of many people invested with power... . Life is becoming harder and harder for common people, especially for those with the fewest social protections.

3. Raising these questions in a very strong way, we are at the same time deeply convinced that we have an opportunity to overcome today's difficulties. What is necessary for this? We believe:

- That we must not delay passing fundamental laws about property, land, socialist enterprise, and taxes; ...these laws should be put in force immediately... .
- That the government must resolutely carry out reforms to establish an integrated competitive market of consumer goods, means of production, investments, currency and manpower... .
- That emergency actions must be taken to end our country's crisis. The government's measures are correct, but they are not enough. The main problem is devaluation of the ruble. But its stability should be achieved neither at the expense of the population nor through various confiscations... . Now is the time to give real freedom to the farmers... .
- We must satisfy the consumer market, using not only government, but also cooperative and individual sectors. Finally, it is necessary to learn how to live within our means, to reduce unnecessary or minor expenses, including superfluous investments, the production of useless things, unreasonably high military expenses and costly international obligations... .
- That we must turn our war of words against crime into a real struggle against it, especially against organized crime, corruption, and sabotage; strengthening law observance, we should create at the same time all necessary legal, political, and administrative facilities, and increase the accountability and professionalism of the police.

4. We call upon people's deputies and voters to speed up perestroika, and to do their best to make peace and friendship among ethnic groups become a norm of communication. Every nation, including the smallest, has a right to the independent development of its original culture, its language, and its economy... .

We are strongly against measures that would divide the workers and collective farmers, on the one hand, and intellectuals, on the other... .

We definitely insist on overcoming... disdainful attitudes toward science.

The achievements of fundamental science are an integral part of our national cultural values, its contribution to world culture, and a guarantee of our further development among great nations.

5. Priority should be given to the problem of reliable, effective nature

conservation. It should become an integral part of planning in economics and scientific activities on all levels... .

6. We must be vigilant toward any attempts to stop or reverse democratization. Perestroika and real socialism are unthinkable without democracy. We must strengthen glasnost, which is becoming firmly established right now. We must defend freedom of the press from incessant attacks... . We must complete the process we have started and make human rights in our country correspond to the highest international criteria, as well as to the international agreements the Soviet Union has signed... . We are very worried about those who are working hard at the bottom to keep and develop our culture: school teachers, physicians, librarians, and curators. A nation that does not care for its spiritual wealth cannot expect a great future.

7. An open dispute is going on in the country on Article 6 of the Constitution [which guarantees the leading role of the Communist Party in Soviet politics]. We are convinced that the existence of this article in the constitution is unjustified. The role of a party in a society must be determined not by an article of law, but by its ability to gain authority and to be in the political vanguard. It is evident that the development of democracy will raise questions of the legalization of public organizations that are considered informal today, as well as the problem of future political parties. We believe that all these problems should be solved in time, while examining the draft of a new constitution... .

The necessity of reasonable compromises at any time, and especially now, is evident. But the solution of certain political, social, and economical issues allows neither half measures nor gradualness. We cannot get over a gulf in two jumps; we must make one. Our country should become a modern civilized society, and we cannot possibly blame ideology or national character as reasons for barbarous backwardness and underdevelopment.

Note from author: Definitions of "left" and "right" wings have drastically changed in Russia since ca. 1991. At the present time communists (including Stalinist-type, orthodox etremists) and "national-patriots" (including nazi-type chauvinists) are called leftists, while democrats who criticize Yeltsyn and the government for various deviations from democracy are called right-wingers.

Are the Discoveries of Lobachevsky Paying Off?[a]

When my request to meet the scientist was answered immediately, I could not help feeling grateful, because I knew that USSR People's Deputy, Director of the N. N. Semenov Institute of Chemical Physics of the USSR Academy of Sciences, Lenin Prize winner, Academician V. I. Gol'danskiĭ is a very busy man. It was possible to meet and talk only late in the evening.—Viola Egikova

Vitaliĭ Iosifovich, recently, much has been said and written about the fact that our science is lagging behind, so it is certainly not surprising if this topic arises today, on the holiday celebrating Soviet science.

I agree. But let us first come to an agreement about what is science. This is how it is defined in the most recent edition of the Soviet encyclopaedic dictionary: "The sphere of human endeavor whose function is the elaboration and theoretical systematization of objective knowledge about reality; one of the forms of social consciousness; it includes both the obtaining of new knowledge and its result, the sum of knowledge on which the scientific picture of the world is based; the designation of various branches of scientific knowledge. The immediate goals are the description, explanation, and prediction of the processes and phenomena of reality which form the subject of its study on the basis of the laws it discovers... In the course of historical development it has been transformed into a productive force and the most important social institution, having a significant effect on all spheres of society."

This is, I would say, a somewhat blurt definition, and it omits several of the most important features, which must be emphasized.

[a] Interview by Viola Egikova.

First, it does not say that science is a component of culture. Second, it does not say that the process of making a "product of science," in particular—of fundamental science—this is a creative process, therefore defined as scientific creativity and often contrasted to a trade. Third, science is essentially international. Fourth, science for a long time was considered as not only learning about nature, but also transforming nature, and today this problem has undergone significant changes and can instead be formulated as the gaining knowledge about and preserving nature, stimulating the development and well-being of nature and society, establishing the optimal conditions for their existence.

The attainments of fundamental science are an inalienable part of the national spiritual wealth of every country, every people, and a contribution of universal and historical value to world culture. The primary task of the Academy of Sciences, which cannot be carried out by anyone else, is to make this wealth, this contribution multiply. To attempt to reduce estimates of the value of fundamental science, its expenditures and profits to exceptionally specific economic indicators, to attempt to apply such criteria to the creativity of Lomonosov, Lobachevsky, Pavlov, Vernadskiĭ, Kolmogorov is almost as senseless as trying to do the same for Pushkin and Tolstoy, Mussorgsky and Tchaikovsky.

Nevertheless, we have gotten used to treating science in just that way.

And this, in particular, is one of the reasons for the very inefficient use of scientific work in practice, the transferring to science of the burden of what is appropriate for technology. I and Academician Yu. A. Osip'yan have written about this in detail in *Pravda* a year ago.

This, incidentally, is also one of the manifestations of the scornful, I would even say disgraceful, treatment of science which we see today. If we consider who today in society is constantly under fire, we must name first of all the mass media, and secondly science. It is typical that in both cases the people doing the firing are the same. Upon examination, it turns out that the mass media and scientists today have common enemies. As a rule, they are ardent opponents of the intelligentsia and intellect in general, and of every creative approach to the world surrounding us, to the freedom of creativity, and creativity as such... .

It must be said that such attitudes, unfortunately, find a broad response. Alas, one of our sad traditions is dislike of intelligentsia. In prerevolutionary times this attitude was, as is well known, shown to students ("overeducated") and it was characteristic of the lumpen strata of society.

After the revolution the intelligentsia from the start was counted among the most untrustworthy elements of society. The terms "working intelligentsia" and "popular intelligentsia" (which implies "nonworking" and "antipopular") appeared. For a long time the intelligentsia was contrasted to the "workers." The impression was created that mental work was something that shamed rather than elevated a person, and so the intelligentsia was allotted the role of a layer in society on the background of working class—possessor of hegemony and the peasantry. There was talk of "worker-peasant power," though in fact this power was promptly transformed into power of a quite narrow bureaucratic layer. The peasantry was subjected to depeasantization and mass annihilation. As far as the workers were concerned, they also essentially were only referred to as masters of their own fate; in reality they were not.

Nevertheless, power was referred to as "workers' and peasants' power," and the intelligentsia was simply not considered. One can read crude statements addressed to the intelligentsia even by members of our first government, although it is often emphasized that this government was the most intelligent at those times... .

What happened in the following years is also well known; in rooting out the remnants of monarchical, bourgeois, democratic, religious attitudes, entire layers of society were cut down. This was when science was dealt the first shattering blow. In the 1920s and 1930s the social sciences were completely castrated; Chayanov and Kondrat'ev were annihilated, they attempted to do the same with Tarle... . Then came the defamation of biology, the study of the great works of the "Luminary of all sciences" [Stalin—*transl.*] in linguistics and economics. A serious threat hung over physics, where quantum mechanics and relativity theory were considered particularly harmful. Only the creation of the Soviet atomic bomb removed this threat. Chemistry remained in the zone of intensive firing for several more years. The war against the "bourgeois pseudoscience," cybernetics, held back the development of information theory for decades in our country.

The academy recently restored the membership of a great number of famous names, nearly all of them posthumously. Can all this possibly have occured with no consequences for the development of science?...

Of course, the sword of Damocles—the threat of physical punishment—is no longer hanging over us today. But again a dangerous tendency is developing in the public idea of the "image of the enemy" in the face of science. What is it not blamed for?! And, sadly, the accusations sometimes resound even at the highest levels, to very important audiences; one need only recall some of the speeches at the Congress of People's Deputies. They make one's hair stand on end. It has become nearly the rule in good society to speak denigratingly of science, to edifyingly suggest what course it should take. As a rule, the people making these judgments are quite far from science and from understanding the conditions under which it exists.

Incidentally, they casually suggest that scientific researchers "earn too much." Let us see what "too much" is. Say, a college graduate, having studied for five years, starting at an academic institute receives about 150 roubles a month; this is considered pretty good. Two hundred roubles for a more advanced coworker—hardly an enviable salary! And this is 2–2.5 times less than a trolley driver who needs only six months of training... .

In the United States a beginning scientific researcher receives about $25,000, $2000 a month. Comparing this to 200 rubles at the current rate of exchange, there is a difference of a factor of 60! Even the maximum salary in fundamental science here, which is 1000 or 1200 rubles, cannot be compared with what an established scientist in the West earns, and moreover such salaries are received by about one person in a thousand.

Scientific researchers do not live in a void. They do not make enough money to take care of elementary needs, and people try to obtain auxiliary contracts or to earn additional money in cooperatives, so they end up spending less time on creative work, sacrificing it for trade. All this damages fundamental science. The primogeniture is sold for a mess of pottage. And there is already an alarming symptom: the sizes of the scientific journals are shrinking, the number of interesting articles in them is decreasing, fewer and fewer young coworkers come to work in the library; I know this from my own observations. However I cannot condemn them... .

But if only all our troubles were the material conditions of everyday life! Technical equipment plays an extremely important role in science. And here there is a no-win situation compared not only with the so-called developed countries, but also with those countries which have experienced the "new economic miracle"—like South Korea, Thailand, Singapore, Taiwan, and the Philippines.

All this in conjunction with the disdain of science cultivated in society can have a very negative effect; it will stimulate a "brain drain." The dramatic consequences to which this could lead can be judged from, for example, the case of Europe in the 1930s, when it lost its scientists. The United States gained them. Incidentally, now the United States is obtaining great scientists from Asian countries. The result? Just look at how many Nobel Prizes in science have been awarded to American scientists in the postwar years: about 80! And we have six.

But the point is not comparison of the number of prizes; in the end it is no secret that the work of our own and foreign scientists is not always measured with the same meter stick; we are undervalued. Another important fact is that of the six Nobel Prizes awarded to our scientists, five were awarded for prewar work, and one (Basov and Prokhorov) for work in the early 1950s. The importance of the achievements in fundamental science made by us at the worldwide level has fallen and continues to fall... .

The prestige of our scientific work is certainly falling also because the word "learned" has been cheapened; this term is now used for anyone working on a project, at a branch institute, in the Construction Bureau...

Landau did not like the word "learned" at all. He used to say, "There is a learned cat (in Pushkin's Ruslan and Lioudmila), and we are scientific workers." It is simply immodest for many to call themselves "learned." This word has really been cheapened. We have been successful in producing a large number of so-called "scientists" as we have been in the smelting of cast iron and steel, in the production of tractors. The statistic of 1.5 million scientists is nonsense! This is just like counting all 10,000 members of the Writers' Union as real writers, while the genuine writers among them may at best amount to one percent of this number.

Today a great deal is being said about the danger of mass culture, which is created by mediocricity and produces mediocricity. But exactly as the existence of thousands of pseudowriters devoid of talent does not justify the snubbing of any kind of writers labor, so the helplessness, blunders, or the cupidity and the striving to please the authorities, quite often manifested in the near-by of science, does not justify the condemnation of science itself.

But is it possible for science to get rid of these fortuitous people? The conditions under which scientific collectives operate do not permit this, do they?

That is quite true. It is absurdly difficult to get rid of mediocrities. Even if such a pseudo-scientist is transferred to some division together with his salary, no one wants to take him. He has not been certified, and according to all the rules could be let go, but he writes and writes to various departments. They form commissions, and the commissions earnestly advise you, "Don't get mixed up with him, it will cost you more dearly!"

A possibility might be, for example, to set up a system of contracts; we are already thinking about this at the institute. It makes the certification unnecessary; anyhow, from experience we know it doesn't work. It would make more sense to make a contract with a researcher for some defined work within a defined time period, which could be extended or not. If the contract is successfully completed and the topic of research exhausted, another could be put into effect. At present it is very difficult to move a person from one topic to another. People try to avoid changing the topic they are used to working on even when it is only they who are interested in it. Thus a topic can become worthless ballast for an institute.

Is the researcher himself always the guilty one? If he could painlessly offer his services to another scientific center, how many lives would have turned out differently...

Of course. This, if you wish, is yet another factor which hinders the development of science. The residence registration system, which deprives scientific workers of the freedom of movement within their own country, is an absurdity in a civilized world, but a necessity under the conditions of the colossal inequality in the standard of living in different regions of the country. Now, of course, this inequality is being smoothed out, but unfortunately not because the level of the poorly paid is being raised, but because the situation is getting worse even in regions which were formerly better off than others. One sometimes gets the impression that the most popular and understood ideal is becoming "equality in poverty."

In the United States, for example, it is normal for a scientist to move from one university to another. In Germany, as a rule, even a talented person after defending what we call the doctoral dissertation is not made a professor in the same institute. There it is considered very important that there be mutual fertilization of different scientific centers and schools, which makes it possible, from their point of view, to avoid one-sidedness in research. Clearly, if such moves worsened the standard of living of scientists, no one would agree to this system. We, unfortunately, cannot yet realize such "delocalization" of scientists.

But the contract system would be a first step in this direction. Will it be introduced into the academy or is the idea being developed only hypothetically?

Each institute must decide this itself. Today many rights are being delegated to academic subdivisions, in particular, each institute henceforth will work out its own charter. As far as contracts are concerned,.... Here, there are two dangers. One comes from the side of the central agencies, including the trade unions: it is possible there will be attempts to unify the situation. The second comes from the scientific collectives themselves. In particular, I think that we have overdone the introduction of the election system in the institutes.

My words might seem outdated to some, but I am convinced that the elections for director should be confined to reasonable limits. I know of cases in our institute where, for example, the head of a laboratory made some comments about poor work to his colleagues, and they said to him: "Well, as soon as your reelection comes up we will repay you!" And this is not such an unusual story. This is why, in particular, to avoid this to some extent we decided to elect the heads of laboratories which are parts of the divisions of the institute not in the laboratories themselves, but in the whole divisions. In this case there is a larger probability that the heads will be elected for their real scientific merits, and not according to the principle "you to me, I to you." Otherwise, the contract system loses meaning even before it is born... .

What other new innovations are expected at academic institutes?

In connection with this I would like to express a wish. I would like to get rid of the need of measuring the achievements of fundamental science by short-term results. You know, if such results had been demanded from Lobachevsky, he would have died without them... .

It seems that physicists have the joke: "We will build a synchrophasotron and...

...discover a new particle by such-and-such a date." A synchrophasotron can be built, but the undertaking to discover a new particle at it... . However the synchrophasotron was built at Berkeley specially for discovering the antiproton. And the antiproton was discovered. It was a unique case, where the Nobel Prize was literally "ordered" beforehand. But this is an exception; after all, a discovery is called a discovery because it is unpredictable.

We had a great biochemist, Jakub Oskarovich Parnas, who, incidentally, was also killed in the purges of the NKVD (in the 1940s). It is said that he used to say: "Before I became a Soviet scientist (in 1939, when Eastern Poland was occupied by Red Army) I could never understand, how it was possible to 'plan science,' but then I realized: one simply has to write in the plan what has already been done. And then the plan will always be fulfilled."

But I would like to get rid of this formalism in science—and I repeat, I have fundamental science in mind. It makes it hopeful that the leadership of the Academy of Sciences will pay more and more attention to these questions, to greater understanding of the fact that the academy is a unique institution which must be concerned with the accumulation of knowledge, the expansion of the national wealth.

Incidentally, regarding the leadership of the academy, elections will be held there in a week. Will there be anything special about them?

Yes. The first elections of the leadership of the academy will take place on an alternate basis. Before there was usually one candidate per vacancy, each of which moreover had to pass through figurative 20 approvals. This was essentially appointment.

There will be another special feature that I would like to mention. I have been a member of the academy for 28 years (counting from the time I was elected a Corresponding Member) and I must say that the last academy elections in 1987 were the first to inspire some optimism in me. Thanks to the new rules opening up vacancies, the academy has become considerably younger, and replenished with good scientists. I was happy that in spite of everything, our country still has talented people. It is a shame when science has to develop not owing to something, but in spite of it.

Now a question to you as a People's Deputy: when will science finally obtain the support of lawmakers? It seems that science and education are still being avoided in the Congress and by the sessions.

It is possible, of course, to write a law about science now. But, you understand, the question is the following: where should the fire brigade go first if the fire is all around? I also spoke at the Congress and can now say that I do not envy the President of the country, and I think that his course will be difficult; it is necessary to help him all we can. I think that we need to make more use of worldwide experience, after all, various countries have had to recover from a crisis. It is also very important that the entire world support perestroika.

As far as a law about science is concerned... After all, science does not exist in a vacuum. We cannot fundamentally change the situation in science without changing the situation in the country as a whole, without carrying out a radical economic reform.

However, we have not had a conversation befitting a holiday, so I would like to end by saying that I congratulate my colleagues on Soviet Science Day! And I add what I said at the beginning: I would like for us on this day to think about those burdens which exist in the lives of scientific researchers, of the difficulties which science is facing... .

On the Basic Principles of the Activities of the Academical Institutes

Speech given at the discussion after the report of the Vice-President, Vladimir N. Kudryavtsev, at the Annual General Assembly of the USSR Academy of Sciences, March 1990.

In order not to take the floor several times, I want to speak directly on all the documents under discussion in our meeting: the Charter of the USSR Academy of Sciences, the status of the scientific research institute, and the material, not yet reported on but available to us, on the restoration of membership in the USSR Academy of Sciences.

Let me begin with the charter. I recall part of item 35 of the present charter: "Actual members and corresponding members of the USSR Academy of Sciences cease to be members of the USSR Academy of Sciences in the event of loss of citizenship of the USSR." This paragraph was added several years ago. It contains two more paragraphs before this supplement: "Actual members and corresponding members of the USSR Academy of Sciences can be deprived of this title by decree of the General Assembly of the USSR Academy of Sciences if their activity is directed at harming the Union of Soviet Socialist Republics." And further: "During the expulsion procedure... the order of voting is the same as for elections."

It seems to me that item 35 should be completely eliminated. In connection with this, let me recall one case, which is perhaps apocryphal. One day Academician Muskhelishvili was asked whether or not he was a former prince. Nikolaĭ Ivanovich answered, "How can I be a former prince? After all, it is not possible to be a former poodle!" This answer applies fully to us: how is it possible to be a former Academician? It is senseless to talk about this. In any case, the first paragraph of item 35 which I read is in complete contradiction with the fact that we are gathered here today to elect, actually, to restore many members to the academy who were expelled from it in connection with losing their USSR citizenship. I therefore think it is logical to simply throw out item 35, or at least its first paragraph.

I think it would be appropriate to restore, from the first Soviet Charter of the USSR Academy of Sciences (I do not have the exact text here) the statement of the role of the Academy of Sciences as the authoritative, sovereign, sole censor of its own publications. The academy should have this right, which was granted several years ago to the editors of literary journals, because there is much that is senseless in the procedure of taking scientific materials through official censorship bodies. The authors of the materials and the institutes which submit them for publication are the ones who know best whether or not these materials contain any secrets. Meanwhile, the additional review and approval of scientific materials for publication is being entrusted to a hierarchy of persons who have less and less understanding of this question as one moves upward through this hierarchy.

Now I turn to the statement about an institute.

- I propose that to item 3 we add: "An institute independently resolves, within the material resources available to it, the economic and organizational questions of its activity."
- Item 4.5: an institute "exchanges information and contacts with scientific institutions and scientists of foreign countries in accordance with the established order." What is this "established order"? If this means the often completely servile dependence on official censorship bodies, then this item will need to be seriously worked on and perhaps changed fundamentally.
- Item 26: the term of office of the director is not indicated. It is apparently five years and perhaps (this still needs to be decided) no more than two terms are allowed. This must be discussed.
- Item 31: it is not clear what is meant by the representation of social organizations in a scientific council. In the past this was stated precisely: representatives of a party committee, a local trade-union committee, the Young Communist League, a council of young scientists. There is still the council of employees, the existence of which in an academic institute seems to me to be completely senseless. I think that the solution of the problem of the representation of social organizations in a scientific council must be left to the institutes themselves (for each to do as it sees fit) and their scientific councils, which are elected on a personal basis.

Finally, let me consider the specific proposals for restoring membership in the Academy of Sciences. In the document we are discussing there is no mention of physicist and Corresponding Member of the USSR Academy of Sciences Aleksander Filippovich Val'ter. He was one of the first three students of Nikolaĭ N. Semenov. He was elected in 1933 by the Division of Physical and Mathematical Sciences, arrested in 1937, and died at the end of 1941. According to the information we have, he was thrown by criminals from a train which was carrying prisoners. I propose that we add Val'ter to the list. The question should also be raised of restoring membership in the academy to scientists who lost their citizenship in connection with going abroad in recent years. Corresponding Member of the USSR Academy of Sciences Veniamin G. Levich has already died, Corresponding Member of the USSR Academy of Sciences Serguei M. Polikanov is alive and is now in Germany. Finally, there is a case where a member of the academy is alive and well in the USSR: I have in mind the so-called "fellow-traveler (of Molotov and his "antiparty" group—in 1957) Shepilov," Corresponding Member of the USSR Academy of Sciences Dmitrii T. Shepilov.

Gurii I. Marchuk asked those present to concentrate their attention on three corrections to the Charter of the USSR Academy of Sciences which were brought up for discussion by the General Assembly. All the propositions, including the exclusion from the charter of the paragraph about membership in the Academy, merit serious study, so that no contradictions arise with other statements in the charter. Marchuk proposed that this work be assigned to a special commission formed by the General Assembly.

ART, MUSIC, LITERATURE, AND THE MEDIA

The title of the third section of Part I speaks for itself—I was often interviewed by different newspapers and journals on non-scientific topics and selected here several typical materials of that kind.

I clearly understand that numerous names and titles mentioned here which are well-known to Soviet audiences can sound strange to foreign readers, but I am sure that they are worth introducing.

Using Both Eyes

Editor's Introduction: During exam time, a good student is often said to have "his nose in a book."

Schoolmate salutes these students, but would like to remind them of the old proverb: "A change of activity is the best rest."

The same is true of books... .

A banal fact: Physicists during their leisure time are drawn to poetry. But I also read books about scientists and science—popular-science books, and fiction, anything... .

Some chapters of the new novel by Daniil Danin on Niels Bohr have been published in recent issues of the magazine *Chemistry and Life*. These chapters discuss, in particular, quantum mechanics. To many it seems that it is nearly impossible to understand this science from the viewpoint of common sense. It is said that one professor, when starting to teach a course on quantum mechanics, first warned the students not, for heaven's sake, to try to understand anything, because they would only get confused. Danin has managed to do the impossible. He has succeeded in finding ordinary words, images, comparisons, and analogies to describe the meaning of something which would seem to directly contradict common sense. In this regard his book *The Inevitability of a Strange World* is extremely interesting.

Now a few words about this type of literature, which may arbitrarily be termed "science-art" literature. These are books about scientists, not necessarily ones who have actually existed, and about life in science.

One of the favorite books of this type from my childhood is *Microbe Hunters* by Paul de Kruif. It seems to me that its strength is that it does not simply make us interested in some specific scientific problem, but to a large degree it makes us interested in the entire world of science and its practioners, with all their weaknesses and strengths, evoking admiration and envy.

I would include in the same category *Arrowsmith* of Sinclair Lewis and the now nearly forgotten but remarkable book *In a Tiny Laboratory* by Nigel Balchin. The distinguishing feature of these books is their ability to inspire in the reader—especially a young one!—the desire to become a part of this world.

There is another impressive writer, Daniil Granin. I find it extremely interesting to see how he reveals in his books the psychology of a person working in science.

"This is correct, and curious, and so unexpected! But true... ." I find in these words feelings which previously were not realized. And one begins to see one's work, one's colleagues, friends, students in a different way... .

To have a stereoscopic view, we need to look at something from at least two sides. Put your hand over one eye and you see that everything becomes flat.

Depths of Life Reflected on the Screen

"Humanity is bewitched by the unrestrainable development of science, and only art can return it to reality." I often recall these words of George Bernard Shaw when I think about the cinema or, in particular, about popular-science films. Since I work at the Institute of Chemical Physics of the USSR Academy of Sciences I often run into cinematographers working in this field. It is sufficient to recall only recent films like "Scratches on Ice," "An Explosion Builds Molecules," "History of the Fire Experiment," and "First There Was a Contradiction," which were concerned with various scientific problems being worked on at our institute.

This gave me a behind-the-scenes look at science cinema. I would like to share my observations here in *Sovetskaya kultura*. After all, we often treat the concept of "culture" too narrowly, using it to refer only to literature and art. However, science—chemistry, physics, biology, philosophy...—is also a part of the spiritual culture of society. Therefore, *Sovetskaya kultura* can to some degree also be considered our newspaper.

It is no secret that viewers go to art films on purpose, but they encounter popular-science films accidentally—these films are shown as an entry to the main program. And it is not only poor advertising that is responsible for this but the makers of the films themselves, who often do not succeed in presenting scientific work in a lively and captivating way. Often instead of a film one gets a kind of illustrated information: a general view of a building, a sign showing the name of the institute, a superficial survey of this and that, an interview with the director, glimpses of people in white coats, apparatus, stands, tables. A strange thing occurs: the film whose star was supposed to be a scientific problem remains without any problem.

In my opinion, much of the reason for this is that script writers working in popular-science film often do not have their permanent favourite field, and sometimes have a quite weak understanding of the main research areas in modern science. One day they write about psychology, the next about genetic engineering, the following one about archaeology. This leads to superficial judgments and indifferent treatment of the scientific material. It might make sense for studios making popular-science films to collaborate more closely with large scientific centers, so as to be in the middle of recent scientific achievements and become more closely acquainted with the world of scientists.

It should also be remembered that scientific concepts have the right to be regarded as aesthetic in addition to informative. The equation of quantum mechanics can certainly be termed beautiful, just like purely technological constructions: a building, a space satellite, or a proton accelerator, as well as a piece of music, people's emotions, a lake in the mountains. After all, in the end beauty is perfection. The great discovery of Einstein is perfect and therefore beautiful. The Periodic Table of the Elements of Mendeleev is also extremely beautiful. It may be more difficult to see this beauty in science, but it is certainly there. And the cinematographers must see it and be able to convey the sense of it.

Another important factor is how popular-science films acquire their (just their, as opposed to accidental) viewers. Here the All-Union Society Knowledge (Znanie) can give some help to cinematographers, because it is the largest organization of our intelligentsia for dissemination of all kinds of knowledge. It can help to ensure that popular-science films reach an interested audience in all the republics, regions, and provinces, and it can distribute explanatory materials literally everywhere in the country.

There is another factor: in addition to specific, narrowly specialized scientific research, there are important general problems in science which have yet to be exposed in film. Examples are the relation between science and ethics, the personal interests of the scientist as opposed to the interests of the state, society, and the scientific collective. I am speaking not only of conflicts among different representatives of science, but also of internal conflicts within the scientist himself, for example, when he is faced with the dilemma of choosing to work on a problem of relatively small scope but with a specific, rapid payoff, or on long-term fundamental research where often not only the schedule for completion but also the nature of the final result are unknown.

Another example is the degree and nature of the risk allowed in medical research. Everyone has heard of the transplantation of various organs in human beings, including the heart, and the associated problems of tissue compatibility. What experiments here are permissible from the viewpoint of ethics and morals? It is one thing for the researcher to risk his own health. But what if the research involves risk to the lives of other people, and if these risks are not taken the possibility of helping humanity as a whole in the future is lost?

The comparative analysis of predictions about how science would develop and what actually happened is also an area rich in topics which have yet to be dealt with in film.

Science today so indisputably dominates the minds of people that some even try to provide wild superstitions with a seemingly scientific foundation. The exposure of such pseudoscientific superstitions, the modern "fruits of enlightenment," is a very important topic for films.

These topics I have suggested might also interest screen writers who focus on feature films. We viewers can only impatiently await new films. After all, cinema and now particularly television films can be justly regarded as the most nearly all-encompassing art form. Films are of interest for everybody, independent of age. I think that my generation, which grew up along with the development of cinematography and witnessed many of the stages of its growth, was the luckiest of all. We loved the cinema immediately, even before we had begun to understand it: "Chapaev," "The Dowerless [Bespridannitsa]," "Masquerade,"... . Now one can often hears it said that despite the large number of films made each year, their high technical level and skillful direction, no film has reached the level of "Chapaev," none has generated such universal recognition and love. I recently had a chance to listen to the argument that the whole problem is the viewer. It is like a stone. If it is thrown into the water ripples appear, and if it is thrown into honey the surface does not stir. Earlier viewers were different, they were more responsive, more susceptible to the magic of the cinema. But I think that this is just not true. Everything simply came together in "Chapaev": talent and devotion of the directors, the writer, the actors, the unusual fate of the hero. Some powerful attraction caused us to watch this great film again and again with undiminished interest.

Yes, many cherished memories are associated with the cinema. The unforgettable Nina Alisova in "The Dowerless." Now one even thinks with some jealousy of how this play could be adapted to the screen by Eldar Ryazanov in his new film "A Cruel Romance."

Even today the heros of "Masquerade," based on a work by Lermontov and incarnated at the screen by Nikolai Mordvinov, Tamara Makarova, and Sergeĭ Gerasimov, stand before me as living people. Although the passions depicted in the film are far from our daily cares, the drama is completely captivating because it concerns the "moral law inside ourselves." There is in Gerasimov's film some exceptional resonance with the poetry of Lermontov. I saw this film during the first months of the war, and I did not then know that fate would allow me to meet with and form an enduring, now 30-year-long friendship with this remarkable couple, Sergeĭ Apollinarievich Gerasimov and Tamara Fedorovna Makarova, that there would be frequent meetings and long heart-to-heart discussions. Sergeĭ Apollinarievich possesses a great store of wisdom in addition to his cinematographical talents. He is an Academician of the USSR Academy of Pedagogical Sciences,[1] essentially because of the encyclopedic nature of his knowledge and his skill in the original, lively presentation of his views on any important problem, including problems in science.

There are topics which touch the deepest chords of the human soul. They make us think about the price, the moral price of human life and happiness, about the meaning of existence.

War is one such topic. The memory of the Great Patriotic War will always remain an unquiet pain in the heart of our people. We therefore all react keenly to any film about war. Anything false immediately strikes the eye, but in return true pictures remain long in the memory. I remember how upsetting the films "The Cranes are Flying," "The House in Which I Live," and "Ballad of a Soldier" were. And how the same feeling was produced by the film "And the Dawns Here are Quiet"... . Speaking frankly, I do not like, and am afraid to go to, long films which are in two series. Their length is usually unjustified. The camera lingers too long on the motion of a city bus, a murmuring brook, or a simmering teapot. However, none of this applies to the picture of Stanislav Rostotskiĭ—you watch it holding your breath.

The last war. This is both a sacred memory and a warning. Now, when there are so many lethal weapons and the threat of atomic war actually hangs over mankind, the antiwar topic has become particularly important in art. Do you remember the old film of Stanley Kramer from nearly 30 years ago "On the Beach" about the tragic destruction of the earth in a nuclear catastrophe? How relevant still today are the words which appear in the final scene, inscribed on a piece of cloth flapping in the wind in an empty, ravaged town: "There is still time, brother."

[1]In 1992 transformed into the Russian Academy of Education.

He Entered Our Lives (Pushkin)

Editor's Note: There is indisputably a Pushkin Phenomenon. The poet has literally entered each home. His matchless poetry and overall creativity, his personality and his fate interest millions. The love for Pushkin and the effect of Pushkin on our multinational Soviet culture are integral components of the development of our society.

On the eve of the 150th anniversary of the poet's death, our editor asked several authors who write for this journal to answer a small questionnaire. These questions were written down 50 years ago by the writer and Pushkin scholar Vikenty V. Veresaev, and the answers were published in the journal *Literaturnyĭ kritik (Literary critic)*. We thought it would be interesting to repeat this questionnaire today. We invite our readers to take part in it.

1. When did you first become acquainted with Pushkin? How did the poet enter your life?

In early childhood. Exactly how I don't remember, but first were the *Folk Tales*. Among my earliest impressions are "The Lay on the Wise Oleg," and how I was sick in bed and my sister read "Dubrovsky" to me aloud. My interest in history led me to "The History of Pugachev" before I first read "Eugene Onegin" and "Boris Godunov." I remember how at the age of 14 I tried to compose in verse a letter from Onegin to Child Harold, which began with the words: "My dear, moĭ dorogoĭ Harold!"

At the age of 21–22 an attraction to Blok and Gumilev temporarily replaced even Pushkin. But soon I again felt, understood, definitively realized how incomparable, how far above everyone else Pushkin was, and his writings, all his letters, all his stirrings of the soul became the most dear to me. During the last 15–17 years I have received much pleasure from the literary evenings devoted to the poet at the Pushkin Memorial Museum on Kropotkinskaya Street.

2. Which works of Pushkin do you like best and why?

Of the poetry perhaps my favorites are "Perfidy", "Desire for Fame," "Autumn," "The Bronze Horseman," "The Little House in Kolomna," and "Count Nulin." And, of course, "Eugene Onegin." And I like almost all of Pushkin's prose, which is so surprising in its clarity and gracious simplicity, and the absence of any attempt to "speak prettily." I will not attempt to explain why I like any particular poem of Pushkin; this is as difficult as explaining why one loves one's wife or children. Of course, an important factor here is the resonance with one's own thoughts or mood at a given stage of life. It is sometimes not even an entire poem, but just a strophe or a line which is striking. For example, I often repeat to myself with melancholy the line "In those days, when all the impressions of existence were new to me."

3. What significance do you attach to the creativity of Pushkin in the history of Russian literature?

Before Pushkin there were sources, while since Pushkin the mighty river of the great Russian literature, so broad and deep, began to flow.

4. What value do you see in the legacy of Pushkin for modern culture?

The value is enormous. It cannot be measured by any statistical analysis of the number of editions or printings, the appearance of operas and films, ballets and songs, television and radio shows based on themes of Pushkin.

All of our contemporary culture would look completely different if there had been no Pushkin in Russia.

5. Has Pushkin's creativity affected you personally? If so, how?

It is difficult to speak of any direct influence of his creativity on the activity of a man working in chemistry, physics, and biology. But Pushkin actively influences the entire course of our thought, our perception of everything around us; he is with us both in joy and in grief, enhancing the joy and sharing the grief. I have often repeated to myself: "If life betrays you..." or "The mirth, now dead, that once was madly bubbling..." and my soul became lighter.

6. Which problems in the study of Pushkin do you consider primary?

The new happy finds like those of Nizhnyĭ Tagil—if we are so fortunate. And, in any case, the passionate propagation of love for Pushkin and following his actions, his thoughts and feelings.

7. What types of editions of Pushkin's works would you like to see produced?

Let the long-awaited three-volume collection of his works finally appear, and then Pushkin will be present in every home. Let the combined efforts of poets, writers, and translators give all peoples on the planet the opportunity of feeling the grandeur and charm of Pushkin, of feeling his nearness, of perceiving him as a contemporary and compatriot. I think that Vladimir Nabokov has made a great contribution to this with his splendid translation of "Eugene Onegin" into English.

On the Road to Recovery

I confess that I cannot be considered a serious expert on the cinema, although I greatly respect this art form. As much as possible I follow the writing on cinema in the press, and I try not to miss outstanding films that they write and speak about.

But once a year for almost an entire month I turn into an inveterate filmgoer, during my vacation. I watch one film after another, and usually sit through every film to the end, because habit dictates that one should not leave the room during the showing. And sometimes I want to leave very badly... .

And then, having become a regular viewer and having experienced the full impact of the current films, I keep asking myself the simple question: who needs this stuff? After all, someone somewhere has presumably decided that "the people need" it. And, in fact, it may be important and interesting to someone else, if not to me.

Of course, it is difficult to decide what everyone needs. We each have our own tastes and predilections, which are quite varied, and depend on individual upbringing and the mood of the moment. Even if an opinion is the view of not just some one person, but of a group of people with common interests and tastes, it cannot pretend to be universal. And here I think it is not possible to do without some sort of institute to measure public opinion: we now are in great need of serious sociological studies which would enable us not only to record the echo of proclaimed slogans, but also to clearly see what are the attitudes and needs of very diverse segments of the population. Today, in the era of glasnost, we are only beginning to study public opinion. There is a great deal of complex work to be done, because as soon as we obtain a more or less complete picture of public opinion, it inevitably becomes clear that the question "what do people need?" has many answers.

But anyhow, if, as the mathematicians say, we count backward, then I think there are only two types of films that certainly should be banned from the Soviet screen: pornography and xenophobia films.

However, we know that there are people who take it upon themselves to definitively judge not only what "people need," but also what "people don't need." They are especially dangerous when they are able to transform their categorical views into orders. It is deplorable that such demands often originate from the viewers themselves: they aggressively bombard newspapers, magazines, television, and radio with letters. They don't like something, so they immediately declare it to be harmful and unnecessary, they demand that no one else have the possibility of watching what they don't like, which is immediately declared to be "seditious." It would seem to be simpler for a person just not to watch something he doesn't like, without making a judgment for everyone. But it doesn't work that way... .

There is much discussion about the need to educate one's taste. Here much depends on the filmmakers themselves, who can now be ruled not by orders from elsewhere, but by their own thoughts, conscience, and experience, and can speak from the screen honestly and impartially. And a lot depends on the critics: on good critics, not clubs supporting the official taste; critics who are motivated, who appeal to the mind and heart of the viewer, who convince the viewer of the correctness of the arguments "for" and "against."

I want to mention several recent films not because of my personal sympathies, but because of their moral and social implications. Here "The Scarecrow [Chuchelo]" should be singled out; it was to some extent a forerunner of perestroika in cinematography. Rolan Bykov was clearly ahead of his time, and it is amazing that "Chuchelo" ever reached the screen during those days! And, of course, there is also "Repentance [Pokayanie]." After seeing it I had a lump in my throat and did not want to talk to anyone. Only afterwards, when I had to discuss the film with acquaintances and friends, did I understand the meaning of some of the complex symbolism and allegories. But the first effect was the deepest shock.

The film "Letters of a Dead Man" made a very strong but oppressive impression. It did not conceal the danger which threatens us all. It should be pointed out that for a long time our films did not deal with the antiwar theme as they should have, in contrast to films abroad, where there are remarkable examples, extremely diverse in genre, like "Doctor Strangelove" and "On the Beach."

Incidentally, here we were also able to watch the American television show "The Day After," which was another film of the fantasy genre about a future nuclear catastrophe. However, I think that this film is not very artistic; rather, it is technical and propagandistic, a guide to action in the case of a nuclear war. It is solid and pedantic, but is not as emotionally affecting as it should be.

On the whole, if we exclude "Repentance," which is in a special category by itself, I think that our cinema has not produced anything comparable to the recent works of literature which have generated so much interest and controversy. The works of Aĭtmatov, Bykov, and Astaf'ev are a bleeding and at the same time a purifying art. This is because the opening of an abcess and revelation of the pain constitute the first step to recovery. I am reminded of the lines of a poem by Robert Rozhdestvensky called "Morning," which was published in a remarkable collection called *Literary Moscow [Literaturnaya Moskva]* in 1956:

> In the end a man perishes
> If he hides his illness.

It is a very important idea; it contains the principal goals and problems of our perestroika.

Therefore, the first step on the road to recovery is to speak openly about the illness, to evaluate what has happened. The next step is the cure. I think that cinema offers great possibilities here, because it is an art form which possesses not only the ability to diagnose, but also to heal. Cinema can thereby help people to orient themselves in the present situation.

Spiritualizing the Mind

Music and music are not alike. Today this formulation, which seems rather obvious, is more and more coming to mean something different: music is in a dissonant state. Its various phenomena—of which there are many—are situated at different poles of life, the everyday spiritual existence of a human being. How did this happen? Why did it happen? It would be nice to understand some of the driving forces, the features of what is happening in this sphere. But the reader must forgive me if in searching for answers I turn to my own "musical experience"—the experience of an unsophisticated person when it comes to art, but a person who cannot imagine life without music.

My favorite composer is Beethoven. My favorite works are his Third, Fifth, Seventh, and Ninth Symphonies, and, in particular, the Allegretto of the Seventh, and in the Third the Funeral March and one part of the Finale. In the latter there is a moment which I love infinitely, it is probably my favorite musical image. It lasts no longer than a minute. I have always marveled at how Beethoven approaches it, gradually, seemingly spiralling into it. And finally, after arriving, he immediately leaves! I think that a composer with a bit less soul than Beethoven could construct an entire composition (or, in any case, many variations) on the basis of this phrase. This approach to a theme and this departure with repetitions is always a great holiday for me. And it seems to me that the lives of the many people who live without such holidays are so poor... .

There is also Grieg. His Concerto for Piano and Orchestra is one of my favorite works. Other favorites of mine since before the war are works of Tchaikowsky (especially the Fifth and Sixth Symphonies and Romeo and Juliet) and of Chopin: his piano miniatures, Concerto for piano and orchestra, and especially his brilliant Sonata in *b* minor. Later my favorite composers came to include also Bach, Mozart, and Rakhmaninov.

The songs from the war years are very dear to me. I did not serve at the front, but the war left indelible traces in my life: the student construction batallion, first on the Karelian isthmus and then near Novgorod, where I was wounded by a "Luftwaffe" bullet, the siege winter in Leningrad and the loss of many friends and relatives, and then evacuation and years of intense work combined with evening studies. However, the war years were the years of our youth. As in the verse of David Samoĭlov, "The fatal Forties, lead, gunpowder...The war makes its way around Russia, and we are so young!" These never-forgotten four years in the end have become the great memory closest to our hearts.

Therefore, I cannot listen to songs of that era calmly. The records of these songs occupy a special place in my large collection (which includes both classical and light music). I don't know which of these songs is dearest to me, the serious "Holy war [Svyashchennaya voĭna]" or the light and lyrical "In a forest at the front [V lesy prifrontovom]." Perhaps my favorite is "Dark night [Temnaya noch']." And there are three songs of the postwar years similar in style to those wartime songs: "The cranes [Zhuravli]" of Yan Frenkel', "The planet shines and turns [Gorit i kruzhitsya planeta]" of Bulat Okudzhava (from the film "Belorussian railroad station [Belorusskiĭ vokzal]"), and the song by David Tukhmanov "Day of victory [Den' pobedy]." I think that Yan Frenkel' is one of the few composers whose songs have a deeply sincere, confidential tone, as though sung to loved ones... . And there is another name which must be mentioned, that of Vladimir Vysotskiĭ. I have many records and tapes of his songs. I particularly like his war cycle: "Song of the Earth [Pesnya o zemle]," "He did not return from the battle [On ne vernulsya iz boya]," and others. It is amazing that someone who remembers the war only from childhood impressions could write such songs. Least of all—which will certainly make all the "certified thinkers" glad—do I like his "bravado" songs. Besides his war songs I like very much the ones dealing with our daily life, our pains and dreams, especially because they contain lyrics with many shades of meaning, which sometimes makes them really piercing.

Having mentioned Vysotskiĭ, I cannot miss here Bulat Okudzhava and Aleksander Galich (the latter once sang in my home), with whom the singer-songwriter tradition began.

I probably first heard Okudzhava in the 1960. How we loved to sing his "Night trolley [Nochnoĭ trolleĭbus]," "Oh war, you false thing, what have you done [Akh, voĭna, chto ty, podlaya, sdelala]," and "The girl is weeping [Devochka plachet]"! They somehow immediately went to the soul. His popularity soared very rapidly after that. Then everything in the singer-songwriter style became immensely popular, became adored. Why?

First, in its best manifestations (which are by no means very rare) it is a real art. Another important factor is that, appearing at the wave created by the Twentieth Party Congress, these songs touched, at the time of the first awakening of our society from the Stalin epoch, perhaps all the problems people were concerned with. They contrasted strongly with the officially condoned type of song, and turned many away from it. Although, the art of such composers as Isaac Dunaevskiĭ and Vasiliĭ Solov'ev-Sedoĭ and the better songs of the war and postwar eras are all unforgettable... .

Anyway, this was something completely different, a new genre which arose at the junction of music and literature. After all, literature was itself on the rise. It is understandable that these singer-songwriter songs were identified with poetry, which was also coming alive in the late 1950s and early 1960s. The memorable issue of *Literaturnaya Moskva* of 1956 signalled this revival... .

If now the question were asked of whether or not the music I listened to during my life led directly to any special thoughts or to science, or if any work of music (or literature or art) inspired my scientific work, I would answer no, I cannot think of any case where this happened. However, it is impossible to organize everything in one's life in files and trace through what lead to what. It is one's general spiritual tuning that is important. Therefore, in order to produce a normal, that is, creative mood, when one wants to live and work, we turn to music. And it, like other forms of art, plays an enormous rule in creating this mood.

From the scientific point of view, I am convinced it is possible to speak of a "reflex to music." This is something unconscious, so to speak, a reaction of creativity in general, of inspiration (not yet understood!) to some external stimulus which excites a certain type of association, which puts us, on the one hand, into a state of deepest contemplation, comprehension, including about ourselves (the past and present) and, on the other hand, into a state of personal activity (an impetus to the future).

Anyway, the fact itself of artistic impressions influencing scientific creativity is indisputable.

One of my favorite books, which I could reread and reread—I am captivated by it as soon as I open it, even though it is a very deep book—is *Everyone Dies for Himself Alone*[1] by Hans Fallada. A wonderful book! The hero (the true hero by name Kwangel) in prison is awaiting execution by Nazi butchers. For some time he shares the room with a professional musician. The latter sings works of Mozart, Bach, and Beethoven to the hero. Fallada gives a remarkable description of how a person formerly unaccustomed to music warms to it, how under the influence of such a simple, unassuming act

[1]"Jeder stirbt für sich allein" (original German title of that book).

his thoughts and his entire mental state are transformed. Kwangel comes to understand the real meaning of music only in this prison cell, shortly before his death. But whereas in the novel the hero is indebted for this "revelation" to an unusual, essentially existential situation, in real life I think that conversely it is diversity, openness, harmonicity of impressions that leads to such revelations. And certainly it is impossible to cultivate harmony in the soul (and then in the character and deeds) of a person while leaving whole areas of his sensitivity untouched. A person should be awakened in early childhood; it is not an accident that lullabies exist. A mother can convey all her feelings simply by words, and the baby understands everything through the tone. Nonetheless, the mother or nanny sing. Nothing can replace this, nothing can take over the role always played by the family in bringing up a child, the family with its habits and atmosphere.

Allow me to recall my family's home. I loved music from a very early age. Many childhood impressions are associated with it. Mama sang, and my sister played the piano. Papa also played the concertina. Sometimes all three played together. For themselves. Our home was full of thick binded books of music, some old, from before the revolution. I remember the words on them: "S. Ya. Yambor Edition. Grand Prix—Rome, 1911." These were works for various instruments, romances—what we would now call popular music. When guests came the house was often filled with music: many of the guests also sang or played an instrument... .

I was nine years old when, while staying at out dacha near Leningrad, my father began taking me to the Kursaal. An orchestra came out on the platform, and it was possible to make requests beforehand. I remember how they played the Valse Triste of Sibelius at my father's request. It was 1932. And after several years my father died. This piece has always reminded me of him, and I listen to it whenever I can... .

What is family music-making? What does it give to people? Not simply aesthetic satisfaction; it enriches and extends the interactions both inside the family and between families. Interrelations and community.

In relation to this, let me tell of one wonderful summer evening a few years ago. My wife and I were in Paris. On one of my free days during this business trip the son of the distinguished German female writer Anna Segers, the eminent French nuclear physicist Pierre Radvanyi, phoned with an offer to show us the city. We spent the whole day going around in his car and walking around in Paris in July, and in the evening we found ourselves in the old Marais district, where right on a lawn some students were playing Italian baroque music for an audience seated on the grass. It was a small chamber ensemble. It was all remarkable, in spite of the fact that the performers were not famous: the warm, gentle evening, crowning the day's impressions, the beautiful Tartini concerto for trumpet and chamber orchestra at the background of summer sunset... .

During my life I have made many different interesting trips around our country and abroad, but this lawn in the Marais district has remained one of my most treasured recollections. Why? Paris, the quiet evening, the charming music in the open air? Of course, all of these. But I think the main thing was the atmosphere of people interconnecting on the basis of art, but in life. They came, saw, listened, sat on the grass, and together with the other listeners were plunged not only into music. Away from the concert hall, formal clothes, and social ritual, this takes its normal course and becomes very necessary. But we do not have enough spontaneous intrusions of art into our lives. It is thought that the institution of wandering actors, musicians, circus people, performing in any suitable place under the sky, came into existence perhaps as an unconscious or perhaps as a conscious response to such a human need.

I have already stated that music is especially important to me for its ability to evoke associations and memories. And since I, like everyone else, was most susceptible at an early age and my tastes and predilections were formed mainly at that time, what I loved in those years is especially close and dear to me now. Just as at a mature age it is more difficult to acquire new friends, also in music it is more difficult to get used to and come to love something new when one is older. Therefore, taking the risk of sounding too conservative, I cannot fail to confess that I find contemporary music, both serious and popular, very hard to accept. What does this imply? A truism is something which for some reason does not take its real meaning and is not realized according to its true sense: that children should study music (as well as literature, history, painting, physics, and so on) in school, that they should become acquainted with all types, styles, genres, and eras of music, including, of course, that of the present. And, of course, that which excites children should not be avoided, even if it appears uninteresting and even bad to us adults. How else is it possible to develop their taste, appreciation of value, and aesthetic immunity?

When my youngest son Andreĭ gives me a list of records to choose from when I want to buy him something, I try to buy some of them. But I do it with distaste, because the list contains so many works of modern Western rock and pop music, which are completely alien to me. I have not even been able to get to like the new Soviet stars in these genres. My wife and I recently heard the rock opera by A. Rybnikov "Juno [Yunona]" and "Maybe [Avos']" in the Moscow Theater of the Leninskiĭ Komsomol. My wife (a piano teacher at the Children's Music School) liked it on the whole, and to me it was not without interest, but the music didn't touch me. Meanwhile, I was not persuaded by the letter—and especially by its tone—of Yuri Bondarev, Vasiliĭ Belov, and Valentin Rasputin published late last year in *Pravda*. This sort of attitude, strong and peremptory, can have only the opposite effect: while desiring to prevent the spread of rock music, they in fact encourage it. Why do I think this? Because a prohibition naturally inspires a spirit of opposition in people. These writers, whom I would describe as engineers or even engineerissima of human souls,[2] who according to the old definition should know how to affect these souls, should understand that methods of "dragging and not letting go" inspire, in accordance with the proverb about the sweetness of forbidden fruits, only the opposite reaction. The principle of "not wasting words where force can be used" has not always held even earlier, and now it is all the more a failure.

Well, rock music is not the chicken pox, but something a bit more serious. It is certainly a mass phenomenon; what do our children express by it, what do they find in it?

It is difficult for me to understand them, but this does not mean that I want to impose my tastes on them. I can only deplore the fact that we often have no understanding of this question and we cannot communicate our perceptions of the same phenomenon to each other. However, maybe they feel just as much regret that I cannot share their enthusiasms with them. I had no such disagreements with my own parents. By the way, my youngest son, who loves rock, also loves serious music (he graduated from music school) and often attends the conservatory; the two interests do not come into conflict. The interest in each is sincere and enthusiastic.

Thinking over why this is so, I arrive at the conclusion that for our children rock is the music of their youth. That is, it is something more than simply music with some particular quality.

Let me recall my own youth. I attended the Leningrad Philharmonic beginning with the last years of elementary school and first years of university. But my interest in the light music of that time developed in parallel; we danced to it and fell in love to it, and passed our very difficult years of youth with it playing in the background. Is this perhaps for today's young people the main charm of music which I cannot understand: the fact that it embodies their youth?

Let me take another look at the past. Everyone remembers how in the prewar years there were attempts to forbid the foxtrot, the tango, jazz; they were branded with various labels, even political. As is well known, this did

[2]The definition of writers as "engineers of human souls" belongs to Maxim Gor'kiĭ, the ironical non-existing word "engineerissima" (plural) is introduced here by myself to emphasize the devotion of three mentioned writers to generalissimo Stalin.

not lead to anything good... . Jazz took its place in the life of the people. I think that today's discussions about whether rock produces or dispels bad instincts are very superficial. Such arguments appear very naive at the end of the twentieth century. I think that bad instincts and, principally, actions are formed by other phenomena more deeply rooted in the life of society. After all, we know that the greatest villains in the history of mankind, in which our century is so rich, listened with sentimental affection to classical music and showed favor to good musicians. It turned out that this favor cast a shadow even on such musicians for many years... .

Regarding today's serious music, it seems to me that it has not yet taken its proper place in our lives. First of all, I think that much of it is beyond the range of perception of the nonprofessional public. I, of course, include myself in this group (however I love listening to the ballet music of "Anyuta" by V. Gavrilin). But the freeing of the spirit, the casting-off which is felt today in literature—in the full extremity of the situation unfolding here—is somehow not reflected in music. This is a pity... .

But overall, I think that the role of art is becoming much more important. In regulated society it was certainly dimmed and suppressed, while today this is being overcome by democratization and glasnost.

It seems to me that we in science can only envy what is being done in literature. In any case, when we read (and it is difficult to choose what to start with!) our literary magazines, they inspire us in the full sense of the word. It is as if one is hearing the appeal: "And where are you?" In fact, where are we all?! We must not simply be fans, onlookers of perestroika, but players in the field, driving the ball. It is true that people from science are today doing a great deal in international affairs. It was essentially scientists who made the world understand that war between states—as a method for solving an argument—is simply unacceptable. Science has played an enormous role in producing the new thinking. However, it could today mean much more in the public life of our country.

In addition to many other things, science shares with art the problem of morality. Does, for example, a high potential for scientific creativity always go hand-in-hand with a high level of morality? I can sadly state that it does not. It can be the opposite. It can be something different. Highly organized thought does not itself ensure morality, it can sometimes be completely immoral.

Take, for example, a person of great ambition, a feature which also is not unambiguous. It can stimulate very positive results. Even heroic deeds. But it can also lead to completely unpleasant, even evil results. An unsatisfied ambition, self-conceit, and envy can transform a person. And a love of power can only be an accomplice of evil. I do not know of any example in history where it has stimulated scientific or artistic activity, I cannot think of any creatively active holders of power. These are two mutually opposed phenomena. But for some reason people with the "power-loving" gene like to think of themselves (and want to impress others with the fact) as at least not strangers to art, and a lot of them think that they understand it so well that they can control it.

I will never forget an expression used by Stalin in reference to Hitler in the postwar years (it would be interesting to find it in the old newspapers), indicating a shocking lack of understanding of the human soul, of the soul of art. He uttered the words "the Don Quixote manners of Hitler." Monstrous words. It is difficult to believe that this was seriously read and accepted—this extremely crude incomprehension, apparently organically characteristic of him, of one of the most remarkable and bright images in world literature.

The meaning of art in society and the relation of society to art have different manifestations. Let us recall: Boris Asaf'ev was an Academician, Igor Grabar' was an Academician, and Dmitriĭ Likhachev is an Academician already for a long time. And so on. What about today? There is nothing of the kind. That is, an important part of intellectual activity and the creative life of the country (the study of music, literature, art) is for some reason outside the USSR Academy of Sciences.

This is not surprising. Is it possible to cultivate in a garden which for many years has not been watered, but only weeded, anything but weeds? Can all the original vegetables suddenly begin to grow? Of course not.

Science and dogmatism are completely opposed. Science is creation, and dogmatism is cramming, repetition by rote. But if creativity in science is suppressed (because it is fraught with...), and only repetition or, so to speak, commentary on the Koran is considered as creativity, science will sooner or later degenerate. It is curious that it is even possible to construct a sort of "hierarchy" of degeneracy. In the humanities this has earliest affected history and philosophy and economics. Literature and linguistics were affected later. But the blow sustained by linguistics in 1950 should not be forgotten!

Let us again turn to the figure of Dmitriĭ Likhachev. What a remarkable person! It is extremely gratifying that when in his eighties he was full of energy, and every one of his performances delivered true aesthetic enjoyment. What aesthetics! I deeply regret the fact that one of our remarkable linguists, on whom I had pinned high hopes, Academician Georgiĭ Vladimirovich Stepanov, died in 1986. He was a highly educated, original, and creatively intelligent person, representing the modern science. There are, of course, eminent humanitarians in the Academy, but unreasonably few of them. And what can be said about the study of music? It is not "honored" at all. Why? There are several aspects to this question.

At one time in the USSR Academy of Sciences there was simply a division of social sciences, while now these sciences are organized separately: there is a philosophy division, divisions of economics, history, literature, and language. To which would, for example, Igor' Grabar' be elected today? Last year I was elected to the American Academy of Arts and Sciences (yes, that is what it is called). I know that in different countries there are academies of arts and literature. This means that the structure of our academy must allow for the role of art and the study of art in the actual life of society. For the time being they have fallen somewhere in the cracks, which is related to the undervaluation of spiritual values and is an additional indication and, perhaps, an additional consequence of the fact that for the last 20 years the Academy of Sciences has become strongly technical.

In this quarter of a century the relative role of the fundamental sciences which form the theoretical bases of the natural sciences has diminished significantly. This implies an undervaluation of fundamental science in general. It often happens here that an immediate requirement, a promise, or present needs overshadow the fact that science is the fount of knowledge. There is more worry about the percentage coming from the capital than the capital itself.

Technicalization is a process which has various negative consequences for the future of humanity, which is strongly ruled by the political ambitions of countries and by situations, when the world at any moment can be pushed to the edge of a military catastrophe. By itself technicalization does not contain the misfortunes which are attributed to it. It is impossible to regard science as a panacea for all the ills of humanity, or as the main source of these ills. Meanwhile, if we consider the last 40 years, we have swung from one extreme to another. Forty years ago all our hopes were based on science, and now it is often spoken of with a curse. Ilya Il'f in his *Notebooks [Zapisnye knizhki]* said, "In the nineteenth century it was said that happiness requires only a radio, now we have the radio but the happiness is absent as before." But he did not blame the radio for the fact that humankind had not found happiness.

Now, at the end of the century, science has unprecedented power (with sadness I recall the striking slogan: "We cannot expect favours from Nature, our problem is to take them from her") while there is a deep deficit of spirituality. Why? It seems that the future of humanity will in the end depend not on scientific-technical guarantees, but on how much it, humanity, recognizes the need for the proper proportions, the correspondence between its collective mind and the possibilities which it commands. If mind and morality are separated from technological possibilities, humankind will perish. If they govern them, humankind will survive, and even uncover new reserves which are at present difficult to predict. The point is to spiritualize the mind (as long as it is understood as something rational), and here art has an important role to play. I do not want to prophesy here, I simply want to express the hope that the spiritualized mind will somehow gain the upper hand over the foolhardiness of bare technocracy... .

Disastrous Shift to the Right[a]

Editor's Note, *Inzhenernaya Gazeta:* The transition to a free market is not only causing material losses (owing to price increases) to most of us, it requires considerable mental fortitude. It is necessary to learn to live in a new way. And we are ill-prepared for this. The universal crisis has lowered the moral state of society to an unprecedented level. Much was hoped for from the Fourth Congress of People's Deputies of the USSR. But it can hardly be said that our hopes were fully justified. This was the starting point of the conversation between our correspondent T. Minaev and Academician Vitaliĭ Gol'danskiĭ.

Minaev: *Vitaliĭ Iosifovich, if the Congress took place today and you had to speak again, what would you say?*

Gol'danskiĭ: I would say just what I said in December concerning the resignation of Eduard A. Shevardnadze, which greatly shocked me. The events which followed the Congress confirmed his fears. The alarm he sounded before the Congress was justified. And now the public is dominated by this uneasy mood. The Congress itself evoked this mood, because it demonstrated that the government of the country has made a clear and pronounced shift to the right. And now it is necessary to seriously raise the point that such shift is inadmissible, because it signals the end of perestroika.

It is becoming more and more common to draw historical analogies. The reaction throughout the world to the resignation of Shevardnadze is a vivid reminder of the dismissal of Litvinov in 1939. And the events of mid-January have an obvious analogy with those of 1956, when two conflicts arose almost simultaneously: the military intervention of the Soviet Union in the internal

[a] Interview by T. Minaev.

affairs of Hungary and then the Anglo-Franco-Israeli aggression against Egypt. Finally, what is happening to us now is reminiscent of the events of October 1964, when Khrushchev was thrown out. He was thrown out after he, by his own speeches and actions, had alienated those who had supported him most warmly: the intelligentsia. Everyone in the world today remembers this, and there is no need for comment. I would have spoken of this at the Congress. The only problem was that they would not give me enough time for the speech.

Minaev: *In this critical situation, what role would you assign to spiritual life, which brings so much that is good and happy to life?*

Gol'danskiĭ: It is now fashionable to cite the Bible. In the Scriptures it is written: "The Kingdom of the Lord is inside us." Let me also quote my favourite lines from Ecclesiastes (7.10): "Say not thou, What is the cause that the former days were better than these? for thou dost not enquire wisely concerning this." A bitter and wise statement. It seems to me that the spiritual source must primarily be based on what is inside us. We need to speak a bit less; we have gotten tired of talks, beginning with the constant worry about what is happening around us, and ending with talk from the highest rostrums, up to the President. The human soul should have a holiday at least once in a while.

Minaev: *The last century left us names, while the present one, it seems to me, makes us ponder over the personification of culture. After all, a person who brings us beauty should not be a part of a herd, he or she is a personality.*

Gol'danskiĭ: I think that we are quite familiar with the people who are creating the spiritual values of society. Moreover, we have recently discovered many names we had forgotten or which are completely new. However, as is typical of turbulent periods, we know not only the work of the artist, but also the social and political image of the person, which at times overshadows this work.

In connection with this, I would like to speak not about new names, but about some names long known to us. I was always deeply sympathetic toward the work of Trifonov, with whom I was personally acquainted, and I was also interested in Rasputin. When reading Trifonov, I experienced the joy of recognition, because of his representations of people among whom I live, and when reading Rasputin I experienced the joy of acquiring a new acquaintance, because in "Farewell to Matera [Proshchanie s Materoĭ]" and "The Last Deadline [Posledniĭ srok]" I saw the living images of people with whom I had not become acquainted in life. Today, when there is almost complete polarization, it is clear to me that wherever Trifonov would be we would have complete mutual understanding. However, many of the opinions of Rasputin evoke sad bewilderment in me, and the fact that I consider him to be a very good writer enhances the sadness. Incidentally, let me note that I have received great satisfaction from reading the essays of another very strong writer, Viktor Astaf'ev, about the singer Viktoriya Ivanova, about the film "The Great Waltz [Bol'shoĭ val's]," about the writer Viktor Konetskiĭ, whom I also highly regard. I could hang on every word. Astaf'ev's positions on many questions evoke strong sympathy in me, since they are primarily and fundamentally humane.

I like such excellent writers as Vasil' Bykov, Vyacheslav Kondrat'ev, and Anatoliĭ Pristavkin, with their deep humanism and fine sensitivity. I love Fazil' Iskander, whose humour is close to my sense. When you discover something new of his you involuntarily smile. And this is now perhaps the rarest facial expression to be seen. Recently, one of our Academicians, upon meeting me, shook my hand and asked, "What bad news can you tell me?" A very bitter but quite appropriate joke. A smile is perhaps what we need most.

To continue our conversation, let me say that I cannot at all agree with the idea that we are suffering a crisis of creativity. As far as art forms other than literature are concerned, it is not only our cinemas, for example, that are empty. I even think that ours are fuller than "theirs." And it should not be forgotten that cinema has a lot of competition from television and video, which are in fact having a general effect throughout the country: sometimes people simply don't want to spend the evening away from home. As a rule, I am able to watch new films only when I am in the sanatorium; otherwise I have no time. Of those I saw in the last year or two I liked "Fountain" and "Taxi Blues," and a good earlier one was "Blue Mountains [Golubye gory]." "Fountain" and "Blue Mountains" are brilliant grotesque spoofs of our life. I deliberately did not go to see "To Live Like This Is Impossible [Tak zhit' nel'zya]," because even without it I know that it is impossible to live like this, and it would have been bitter and unpleasant for me to see it.

The theater? I am sure that the shrinking of the audience is connected with the fact that people have to be in the right mood for theater, as for any other art form. This is because theater is a holiday. And to go on holiday—I have already spoken of this—ones needs to be in a holiday mood.

Minaev: *How is it possible today to hold onto the rational, the kind, the eternal which is being lost?*

Gol'danskiĭ: First it is necessary to want to hold onto these things. I am getting the impression that many of our cultural figures—and it is really a shame if this group includes talented people—are continually seeking and creating the image of an enemy, generating hatred, and are ready as a preventative measure to destroy all imaginary enemies.

I am frightened by the medieval level to which we are falling and which is demonstrated by the unprecedented flowering of belief in all sorts of pseudoscience—astrology, chiromancy, all sorts of omens and augeries. Are we strong enough today to keep alive what is rational, kind, and eternal? Some are strong enough, others don't want it, and yet others think that the worse things get, the better they are. I would like it if somehow or other each artist were as good as he is talented. I am sure that these categories must overlap.

Minaev: *Scientists are undoubtedly essentially creative people. If we consider the range of interests going beyond science and extending to art, to social life, then are our scientists similar to their colleagues abroad?*

Gol'danskiĭ: That is an interesting question. Let us begin with music. Among those doing research in the exact sciences, there are more scientists abroad compared to here who can (and love!) to play musical instruments. As far as literature is concerned, I would give the first prize to our scientists. They are probably better acquainted with their native and with foreign literature than their foreign colleagues are. In the area of the visual arts it is hard to say. Appreciation of modern art, which was completely denied to us for a long time, as though it didn't exist, is of course more common in the West.

But it is difficult for me to judge, because to tell the truth I do not follow either the latest "isms" (my preference is mainly for Impressionist painting and extends up to about the 1920s), or rock music, or other supernovae. It is probably an effect of age—I tend to lean to older things.

In general, comparison of the relation between science and art here and in the West is difficult just because our scientists for a long time were living in a sort of half-smothered state. At the very least they were gagged. During all these decades scientists in the West were able to develop in freedom not only in science, but also in thinking in general. And free-thinking is a great word. It is a condition necessary for existence.

Rating TV[a]

Pasyutina: *Which television shows do you like? Which don't you like? Which hosts on the screen appeal to you most?*

Gol'danskiĭ: Unfortunately, I watch television only occasionally, so it is difficult to name any program that I find particularly memorable. I am interested in those related to history. After all, history is very rich and deserves a great deal of attention. Now there is a variety of high-quality programs about history. And it is good that our history is being looked at from various viewpoints, sometimes conflicting ones. Even if one does not agree with the position of the author, the experience is still enriching.

I try to watch news shows, both in the First and Second TV programs, in order to keep up with things, even though the contents always produce negative emotions. In my opinion, it is inadmissible to constantly plunge people into despair; it is necessary to try to find some "ray of light in the dark realm."

Television at present is greatly debased by serials like "Santa Barbara" and "The Rich Also Cry." I think this is simply monstrous. A normal person might watch these shows a couple of times, but to make this a daily activity... ?! Here television demonstrates complete lack of taste and spoils the taste of the viewer. We have many first-class films by real masters, but we almost never see them. I would gladly watch one of the classics like "Masquerade," "Chapaev," or "The Dowerless" by Protazanov. It was great to see "That Hamilton Woman." I think that a series "Masterpieces of International Cinema" would be a good idea.

The TV hosts I like are those who are not indifferent to the viewer and to the information they present. They address the viewer as a good friend, with affection. And this calls forth a response. The weekly show "Summary"

[a] Interview by Evgeniya Pasyutina.

hosted by Evgeniĭ Kiselev is very good. The last installment, devoted to the May events at Ostankino, was really an event. It has explosive force, and they are up in arms against Kiselev.

Effect of a Natural Contraction of Time
(Facetious Attempt at a Mathematical Description of Age Sensation)

A well known phenomenon in physics is the Lorentz contraction of time for bodies moving with relativistic velocities (v) close to the velocity of light (c). It follows from relativity theory that for a space-traveller accomplishing superrapid intergalactic flight, time will pass much faster than for his fellowman remaining at home.

It would seem to the traveller returning to Earth that he has spent in flight some time, T, while for the inhabitants welcoming him home on return to his planet his absence was much longer, namely:

$$t = \frac{T}{\left[1 - \frac{\mathrm{v}^2}{c^2}\right]^{\frac{1}{2}}} \tag{1}$$

However space-travelling is not absolutely necessary for the acceleration of time. Time is found to accelerate by itself—first it goes, then it runs, than it flies, while we get older.

Below are proposed some simple formulas describing this effect of the natural contraction of time which, although long known, has heretofore not been described in a mathematically simple fashion.

If the time, which has passed since our appearance in this world equals t, and is increasing by dt, then the apparent increment of time, dT, will always be shorter than the true increment, dt; in other words—the apparent run of time is always faster than the real one in such a way that:

$$\frac{dT}{dt} = 1 - \frac{t}{t_m} = \frac{\Delta}{t_m} \tag{2}$$

where t_m is the total (natural) duration of our life, and $\Delta = t_m - t$ is the lifetime remaining at time t.

It is easy to see that the apparent progress of time sharply accelerates with aging—e.g. if $t_m = 90$ years, then the time progresses for an 80-year-old twice as fast as for a 70-year-old individual, and for this latter-twice as fast as for 50-year-old. As t approaches t_m, the run of time accelerates to infinity.

One can also easily see by integration of (2) that

$$T = t\left(1 - \frac{1}{2}\frac{t}{t_m}\right) \tag{3}$$

and consequently when t approaches t_m, we feel that life has passed twice as fast as it should have. This effect is undoubtedly the basis of the widespread interest among mature persons in what one might have done with the possibility of starting over with a new life.

As illustrated, the following is offered as a short problem for the readers. The author's present age is 54. It seems to him that the time now runs twice as fast as in those days, when he was 38, and three times faster than in the age described in a very impressive way by a poet[1]: "Here I go, handsome, 22 years old." What is the value of t_m assumed by author for himself, and can he be treated as an optimist or as a pessimist?

Reference

1. V. V. Mayakovskii. "Oblako v shtanakh" (Cloud in trousers), 1916.

PART II. SPEAKING OUT

FOR PEACE AGAINST NUCLEAR THREAT

Since 1977 I have been actively involved in the Pugwash Movement of Scientists. I took part in many Pugwash conferences, workshops and seminars, and in 1987 became the Chairman of the Soviet (now Russian) Pugwash Committee, a Member of the International Pugwash Council and the Executive Committee. Out of this involvement grew various speeches and publications, particularly on the problems of CTB (Comprehensive Test Ban) and NPT (Non-Proliferation Treaty). Section 1 of Part II contains a selection of publications on this topic.

Out of the SDI Impasse[a]

Editor's Note, *Izvestiya:* Five years ago, Ronald Reagan, President of the United States, in a nationally televised address said that he had ordered the creation of a long-term program with the goal of eliminating the threat posed by nuclear missiles. This program costs trillions of dollars and does not increase the safety of the United States. Moreover, the introduction of weapons into outer space would lead to destabilization with unpredictable consequences. The Strategic Defense Initiative (SDI) impedes progress in nuclear-missile disarmament.

Izvestiya asked Academician Vitaliĭ Iosifovich Gol'danskiĭ, chairman of the Soviet Pugwash Committee and vice-chairman of the Committee of Soviet Scientists for Peace, Against the Nuclear Threat, to give some preliminary conclusions and his impressions about the future of the planet taking into account the Star Wars factor.

Guk: *How far has the U.S. advanced in this program during these five years?*

Gol'danskiĭ: I cannot judge about all aspects of the work. If we take the one component about which there has been so much noise, the x-ray laser with nuclear pumping (this is closer to my interests, since I myself once worked on the gamma-ray laser), here the American scientists have different opinions. In Livermore Laboratory itself, which is the main developer of the x-ray laser, the scientists argue constantly. The skeptics think that this ven-

[a] Interview by Sergei Guk.

ture simply amounts to throwing money away on a project which is clearly unrealizable.

A funny thing: when at a specialized international symposium we recently discussed problems related to nuclear testing, I asked one of the people from Livermore if he seriously thought that it was possible to create an x-ray laser system which could intercept many hundreds of missiles. He answered no, we do not think so, but we are carrying out this work to see if the Russians can do it. A great anecdote!

There was other news: the ideas for SDI which were first put forth have undergone very great changes in the direction of trivialization, for example, for the demonstration of rapid success of the missile annihilation program using missiles with completely ordinary, chemical fuel. And when they keep talking over and over again about early deployment, they are primarily pursuing the goal of avoiding having their funding cut off. They try to create the impression that some breakthroughs have been made, while in fact the entire program is at present being revised. If they had stuck to the original project, they would be forced to recognize that the deadlines for realization are constantly being extended.

Guk: *Incidentally, didn't the proponents of SDI from the beginning promise hundred-percent shielding of the population?*

Gol'danskiĭ: No one is thinking about this now. The overwhelming majority of American scientists view SDI in the form first envisioned as technically unrealistic, expensive, and politically harmful. The supporters—employees of the weapons laboratories which are essentially feeding off the SDI—go on with their work. But there is also a third group which one hardly hears about: I would call them eyewashers. They try to present their proposals such that, even though completely unrelated, they appear to have something to do with SDI, and in this way obtain funding from the Pentagon.

Guk: *Allow me to ask a naive question. Let us assume that a hundred-percent shield is created. We assume that in the case of a war only the unshielded part of the planet will suffer. But will the shielded part really be spared?*

Gol'danskiĭ: It is sufficient for one side to explode its reserves of nuclear warheads on its own territory for human civilization to be irrevocably annihilated. The only difference is that such a suicidal action would finish off that side immediately, while the others would have to undergo a long and painful death.

Guk: *There is now apparently a stratum of politicians and military people like during the First World War, when destructive new technology was available, who continue to think in terms of howitzers and airplanes.*

Gol'danskiĭ: Let us take the case of only conventional warfare. The earth is now heavily armed energetically with nuclear power plants. The complete destruction of a million-kilowatt nuclear reactor would lead to short-term radiation damage equivalent to the explosion of a megaton bomb. Its long-term radiation consequences would be equivalent to a 10-megaton bomb. And the total power of the nuclear power plants around the world is about 400 million kilowatts.

Guk: *What does this imply?*

Gol'danskiĭ: The long-term radiation damage is equivalent to that of a 4-thousand-megaton bomb. The total amount of ammunition is now estimated to be 5–10 thousand megatons. Therefore, if the nuclear power plants were destroyed their total radiation damage would be the same as that of a nuclear war on a global scale.

Guk: *Now it is inadmissible to rock the Earth.*

Gol'danskiĭ: Not in any case. For example, we do not even know all the possible long-term consequences of the effect of radiation on life. The difference between the genetic structure of a human being and a gorilla or a chimpanzee is so small that it is difficult to even imagine what could be the result of mutations arising from the long-term, global effects of radiation.

Guk: *Why then is there so much noise and anxiety about this SDI?*

Gol'danskiĭ: It's all a lot like an autocatalytic chemical reaction. The sensation which was immediately produced in the U.S. by SDI also affected us. In fact, any point on a national frontier is always described by two coordinates, the latitude and longitude. SDI threatens to create a third dimension in the danger zone: after all, national sovereignty does not extend into outer space. When we express our anxieties about SDI, this is not at all because it deprives us of the possibility of striking back; after all, we ourselves are proposing to reduce and liquidate any weapons, to create a defense system which is not to be used for offensive purposes. We can be disturbed only by the unpredictability of the consequences of the huge capital expenditures going into military research.

However, the violent reaction by our side created false illusions in the proponents of the SDI program: Aha! If the Russians are afraid, that means it's the right thing to do. It means that for all other things equal, our strike will be stronger... .

Guk: *The supporters of SDI at one time were clinging to the words of Michael Gorbachev, who said that we are also engaged in fundamental research in the same areas as the SDI research in the US. Now they attempt to use these words to justify their support for SDI.*

Gol'danskiĭ: We are carrying out wide-ranging, outer-space research. Declaring it to be SDI-related is a rather stable trend in the West. Quite recently we met with a group of American senators and researchers through the Committee of Soviet Scientists for Peace, Against the Nuclear Threat. There was talk about the laser designed for the Phobos project; there had already been noise about this being a project to introduce a laser into outer space, one element of the Soviet SDI. Academician Sagdeev cited specific numbers, and it turned out that the beam intensity of this laser is 100 million times smaller than required for military purposes. I am therefore not surprised by the noise about the allegedly nearly realized Soviet SDI; this is ordinary propaganda.

Guk: *Therefore, SDI is senseless in all respects, but we continue to speak of it as a threat...*

Gol'danskiĭ: I repeat: it is a threat in the sense of the unpredictability of the results of the scientific research, which can turn into a threat to Earth from the cosmos. Today the crazy expenditure of resources by, for example, the U.S. on SDI is comparable to the striving of a gambler to ruin his rivals and snatch a large sum. The elements of bluffing are present, but in addition, knowing the history of science, I can predict already today that this capital expenditure will lead to some sort of result, and not only a military one.

Guk: *But after all, in this case it is not at all obligatory to carry out scientific research within the framework of SDI.*

Gol'danskiĭ: Of course. In this case the enormous resources put into this do go a long way to solving the problem of domination in outer space. At present no one can say what will be obtained from this. This is a destabilizing feature. The fate of humankind will be entrusted to computers. There is a number: the program running SDI must contain 100 million lines of code. This amount of information is comparable to that in a library of 5000 volumes of 300 pages each. There cannot be a single misprint in any of this. The risk of failure is enormous. But let us forget about SDI for a moment. Science is becoming more and more expensive. A computerization process is going on, and experimental techniques are getting more complicated. Giant projects which are of general interest to all humankind must be international.

Guk: *Do you as a scientist have a plan for outer-space defense which would actually contribute to global security?*

Gol'danskiĭ: I think so, and I shall try to explain my idea. The era of nuclear deterrence, in the initial stage when the U.S. monopoly on nuclear weaponry was broken, had a stabilizing value. Nuclear equilibrium then became a restraint on war. But today, in the era of nuclear power plants, even conventional warfare is becoming impossible. The idea of nuclear deterrence has become outdated. The Soviet program of disarmament for the sake of development reflects the new realities of the era. We would like to arrive at relations based on mutual trust. Peaceful coexistence, the development of all countries without attempts to impose on them from the outside is what we must attain. How do we pass from the stage of prolonged opposition to trust, the basis for such a world order? In my opinion, this must be a long stage, which I refer to as the stage of verificational deterrence. This is deterrence based not on the strength of arms, but on rigorous mutual verification while gradually reducing the level of arms. It is necessary to develop a system for observing the Earth from outer space. It must be possible for both sides to reliably monitor each other, in order to gradually learn to trust each other.

Logic of the Mind vs. the Logic of the Ego [a]

Editor's Note, *Izvestiya:* Technology outstripped human consciousness a long time ago. Today we are beginning to reap the bitter harvest of the scientific-technological revolution: the nuclear threat, habitat destruction, hunger and epidemics, and exhaustion of natural resources. What lessons must be learned by humanity at the end of the twentieth century from these developments? Our correspondent S. Guk asked this question of Academician Vitaliĭ Iosifovich Gol'danskiĭ, chairman of the Soviet Pugwash committee, vice-chairman of the Committee of Soviet Scientists for Peace and Against the Nuclear Threat, and candidate for People's Deputy of the USSR from the Soviet Peace Fund and eight other Soviet public organizations.

Gol'danskiĭ: To put it briefly, it is impossible to go on living this way. If we take ecology, it is time to reject the old idea of humankind's dominance over nature. Even Lenin expressed this idea. But everything changes; today we can no longer speak of humankind's dominance over nature, we must think in terms of the peaceful coexistence of human beings and nature and restraint of humankind's aggressive habitat destruction. We must realize that the disasters we have caused in nature will in the end be disasters for us. We can no longer injure the world around us.

Let us take the military aspects of our existence. We can no longer view warfare as a means of conflict resolution, as an extension of politics by other means. Warfare as a means of "clarifying relationships" must be relegated to the past, and this is true for both nuclear and conventional warfare. With today's technology (atomic power plants and chemical factories), the latter could have consequences just as catastrophic as nuclear war.

The concept of mutual deterrence has become hopelessly outdated. In its time it did certainly act as a brake which prevented the outbreak of war be-

[a] Interview by Sergei Guk.

tween East and West, but now it has become a threat to the existence of humanity. The danger of accidentally starting a conflict is increasing; the fate of humanity is entrusted to computers which can go astray. It is time for this concept to be replaced by another: that of self-deterrence and self-limitation, in nuclear technology, industry, energy, whatever. The logic of egoism must be replaced by the logic of mind.

The transition to a new way of thinking under these conditions is unavoidable, because it is based not only on moral principles (which by itself is also extremely important): the changed conditions of our existence require a new level of consciousness. The present level of scientific and technological progress irreversibly forces us to this.

Scientific and technological progress today is based on three things: nuclear power, exploration of outer space, and information technology. The first makes war between East and West impossible. Two other factors have made the world extremely small. They have transformed our planet into something like an apartment building: everyone now knows what his neighbor is doing. The foundation has been laid for a reliable mutual verification system, and a new factor has arisen in international relations: verificational deterrence. There is not yet mutual trust; it must be replaced by mutual monitoring and observation. These will generate confidence. In my opinion this is how we have to move ahead.

Common human values have now become the priority. If threatening extraterrestrials landed on the earth, we would all unite to defend ourselves against them. During the Second World War the USSR, the United States, and Great Britain fought together against fascism. This short-lived alliance was the prototype for a new, realistic way of thinking. Now the role of the enemy is being played by the ecological situation and the threat of war. We are faced with the problem of the survival of humanity.

Guk: *In other words, the world has entered a new stage of existence in which the former rules no longer hold. Today it is no longer possible to protect one's self from others, instead we must be guided by the seemingly paradoxical idea that if we want to live in safety, we must care about the safety of our neighbor. Military options do not ensure safety, but the opposite. Which of the recent events in the military sphere would you cite as an outgrowth of the new thinking, which leads us to hope that our planet can eventually be safe from the threat of annihilation?*

Gol'danskiĭ: There are certainly signs of such new thinking. The treaty with the United States on the liquidation of ICBMs is a break from the past, a transition from abstract political thinking to actual implementation of new ideas. There is a real possibility that a treaty guaranteeing a 50 percent reduction of Strategic Offensive Weapons (SOW) will be concluded. The path to it has been opened up by the earlier agreement to eliminate an entire class of rockets. We are on the threshold of new steps toward disarmament: the East and West have discussed in Vienna their intention to think about a radical reduction in their armed forces, so that they will have a purely defensive character. There is also a realistic prospect of a nuclear-free world. My optimism is strengthened by measures taken by our side: the reduction of the Soviet armed forces and armaments announced by M. S. Gorbachev at the United Nations, and the publication of data on the numbers of armed forces and weapons by the Warsaw Pact countries and NATO. This is a step toward openness, and it is encouraging.

The development of space technology and informatics makes it more and more meaningless—various restrictions which in fact deprive ourself of necessary and useful information. In the process of democratization and glasnost the people must be given as much information as possible. This is an essential component of the new way of thinking, a pledge of full participation of the people in politics.

Next should come publication of data on our military budget. See what happens otherwise. If we use Soviet sources, our expenditures for weapons in dollars (I am now using the numbers from the London Institute of Strategic Studies, which reprinted our data), we find that from 1976–1985 the USSR military budget was about 23 billion dollars annually. The institute also quotes three other sources: the CIA, the Pentagon, and the United States Committee of the Heads of Staff. According to their calculations, our expenditures are more than 200 billion dollars annually. I think that these numbers are much closer to the truth than the official 23 billion, which is simply ridiculous. This is easy to see. We ourselves speak of parity in weaponry. How could this be achieved by spending 23 billion dollars when the other side is spending 300 billion for this? And the entire world understands this very well. This is why they do not trust us. It is time to state openly our present defense expenditures for the sake of our own prestige and to inspire trust in our transformations.

So, I repeat: there are encouraging factors, there is progress in diminishing the threat of war; humanity is turning on the right path, but it must be clearly seen that this path is a long one.

Guk: *But there are many other facts which make one doubt that humankind has actually come to its senses. Consider: everyone concedes that nuclear war is impossible, but the planet continues to tremble from underground nuclear explosions for testing the very weapon which threatens every living thing with total annihilation. We are told that these tests are needed to "improve" nuclear weapons, and that the latter are needed to preserve parity. How can the striving to preserve parity and the new way of thinking be reconciled?*

Gol'danskiĭ: I think that the principle of parity is outdated and destructive.

Guk: *We are allowing ourselves to get caught up in a prolonged nuclear marathon...*

Gol'danskiĭ: Quite true. We can no longer live beyond our means.

Let's compare: the national income of the United States in 1989 was about a trillion dollars a year, while ours is barely more than 450 billion rubles (according to the official rate of exchange, more than 700 billion dollars). If we continue to preserve military-strategic parity, our defense spending (to call things by their names) must be slightly less than that of the United States. It is sufficient to subtract that amount from the national income to see how much remains at both sides for peaceful purposes. If we continue in this way, our lag behind the world economy will continue to grow. We therefore must strive for parity first of all in peaceful areas of production, while in the military sphere we must adopt the principle of reasonable defense capability. It is pointless to compete militarily in the sense of how many times over can we annihilate the enemy. We need parity in the peaceful economy and not in the military sphere.

Regarding underground explosions. I remind you that from 1958–1961 the United States and the USSR adhered to a private moratorium on all forms of testing of nuclear weapons. Unfortunately and to our shame, it was violated by the USSR. The reason given was that France was continuing to test its weapons. Academician A. D. Sakharov spoke out strongly against this decision. At that time we were very close to a complete ban on all testing, but the chance was unfortunately lost. The treaty between the USSR, the United States, and Great Britain concluded in 1963 imposed restrictions in only three areas (the earth, water, and the atmosphere). There were no significant restrictions imposed on underground testing because at that time we were completely opposed to monitoring, and considered it a form of espionage and a violation of the principle of sovereignty.

My personal view is that the recent unilateral moratorium of the USSR on underground nuclear testing does not weaken us at all. And, as far as I can judge from the reactions of colleagues abroad, this step strengthens the authority of our country and its prestige abroad. I am in favor of a complete ban on nuclear testing. Under present conditions such tests are senseless.

Guk: *Whereas some hope can be seen for disarmament, judging from all the evidence the ecological situation is quite depressing. I remember how 10 years ago our prognosticators were boldly rejecting the pessimistic predictions of Western scientists, and showing, like two and two makes four, that the Earth was capable of feeding everyone, that the richness of the oceans was inexhaustible, and so on. The limit has been reached: the Earth's resources and the oceans are stretched to the limit, the protective ozone layer is disintegrating, and the air, water, and soil are poisoned. It is evident that emergency measures on an international scale are needed, but each country individually behaves like a soldier sent to plunder a city. Is there any real chance that in the little time remaining, which some scientists estimate to be less than 50 years, we will be able to pull back from the edge of the abyss?*

Gol'danskiĭ: I think there is, and anyway I believe in human mind. Scientists are raising the alarm and proposing specific measures to save the environment. A recent example is the Dagomys declaration of the Pugwash council. Scientific achievements have provided the basis for today's industry with its negative side effects. However, thc world's scientists are prepared to provide effective aid to save the environment.

Guk: *First destruction, then aid...*

Gol'danskiĭ: Science only makes discoveries, which can be used either to help or to hurt humankind. It is impossible to stop scientific thought and progress.

I would like to recall the American researcher Rachel Carson, who back in 1962 wrote the remarkable book *The Silent Spring*. This was essentially the first alarm that life on the planet is threatened with destruction. Scientists were the first to warn about the destruction of the ozone layer and the greenhouse effect. We remember the enormous construction projects of communism, which only damaged the ecological balance. Were scientists really the ones responsible? Everything was done by the order of the "great architect" i.e. Joseph Stalin. Scientists were the executors. Anyone who protested disappeared in a prison camp. Throughout the world there were some prophets among scientists who were the first to warn of a calamity, and there were also careerists who were ready to do anything. And there were forced laborers—this was the unfortunate situation of those Soviet scientists who at the time were forced to oppose their own conscience.

The main factor operating against ecology throughout the world is energetics. There is a need to provide for the growing energy requirements of humanity. The pursuit of this goal increases the ecological threat, and this is true for all forms of energy. So it appears that there are two contradictory trends. What is the solution?

What I will now say is not a panacea, but these considerations must be taken into account. By economizing measures the West managed to save itself from an energy crisis. After this the situation in the world oil market changed fundamentally: prices fell, which had—by the way—a fatal effect on our economy.

Now let us take the energy reserves of the planet, the consumption of which increases by about 4% a year. According to the data of West German researchers, for the same rate of growth of consumption the coal reserves are sufficient for 80 years, the oil reserves for 40 years, and the natural gas reserves for 50 years. Now let us imagine that by economizing measures the energy consumption is frozen at today's level. Then the coal reserves may be sufficient for 500 years and more, oil for 80 years, and gas for 150 years.

Guk: *Should we leave the problem for our descendants to solve?*

Gol'danskiĭ: No, what I have in mind that even another 40 years can be of great value for advancing the technology of solar energy and controlled thermonuclear fusion. There are alternatives in the distant future. Although we have not decided what the energy sources will be, the problem of the disposal of excess heat in the planet will remain. The excess heat leads to melting of the polar icecaps, the greenhouse effect, and so on.

For the energy sources available today the contradictions between the interests of ecologists and energy consumption are antagonistic. If we succeed in finding alternative energy sources such as solar energy or controlled thermonuclear fusion, the contradictions will still remain, but they will no longer be antagonistic, they will be resolvable. The main problem is to survive in a satisfactory way until we harness new sources of energy. We therefore must economize on energy, and we must use the experience gained so far.

Guk: *Should we understand the new way of thinking to be just political, ecological, and economic thinking? After all, people do not live only by politics or economics. They have some ideas about each other which are more or less accurate. Without trust between peoples there can be no qualitative change in international relations. What must be done to raise the popular consciousness?*

Gol'danskiĭ: The "image of the enemy" must be dispelled. For this it is necessary first of all to get to know each other better. This is why I attach such great importance to mutual exchange of information, between all sides and not just unilaterally. There is a joke: how can two equally popular politicians be characterized when one is "for us" and the other is "against us"? The first is called "popular," and the second is called "not unpopular." The time for rejecting such a gradation is long overdue. We need sober, stereoscopic vision, not euphoria or excessive suspicion. This requires dealing with each other more often and getting to know each other. We need different points of view, we need to compare many viewpoints, in order to decide which is correct. A great deal in this area has happened both in the USSR and abroad; public opinion polls confirm this.

...When scientists debated the question of the possibility of discovering other civilizations in the universe, one of the probability criteria was the lifetime of a highly developed civilization. And what do we mean by "highly developed"? One which can signal its existence in the universe: the countdown to the moment of its destruction begins from this instant. The duration of the existence of a highly developed civilization has been estimated to be many dozens if not hundreds of thousands of years. Now imagine: for us, inhabitants of the Earth, only several dozen years have passed from the time when we became able to send the first signals out into the universe to tell of our existence. And we are already speaking of the real danger of the destruction of humanity, more precisely, of its self-annihilation. We are responsible not only to ourselves, but also to the universe, if only because of the uniqueness of the civilization we have on this Earth. We must ponder over this.

Elimination of Nuclear Arsenals [a]

The very first resolution of the United Nations, unanimously adopted by the General Assembly in January 1946—less than six months after the bombs were dropped on Japan—pledged "the elimination from national armaments of atomic weapons and of all other major weapons adaptable to mass destruction." [1] During the four decades since then, nuclear disarmament has been the recurrent theme in many UN fora, expressing the strongly held feelings of people in many countries. The huge nuclear arsenals accumulated during that period—in pursuit of greater security—have actually increased the probability of a nuclear war. At the same time, the growing knowledge of the awesome consequences of such a war, even if it could be limited in the number of weapons and type of target, made the avoidance of nuclear war the most vital issue.

The call for the removal of nuclear weapons has been echoed by the leaders of the superpowers. Creating a nuclear-weapon-free world is the declared goal of Mikhail Gorbachev, as part of his new way of thinking about common security. In his speech of January 15, 1986 he proposed a timetable for the complete elimination of nuclear weapons by the year 2000. [2] At the Reykjavik summit of October 1986, he and President Reagan discussed an even shorter period, 10 years, to achieve the goal of a nuclear-weapon-free world. This discussion ended in failure, but nuclear disarmament, down to a zero value, has been placed firmly on the agenda of peace negotiations. China and India are also committed to the creation of a nuclear-weapon-free world, the latter giving the year 2010 as the target date. [3]

At the same time, the opposite view, that nuclear weapons must be permanently retained, has also been forcefully expressed. Originally, nuclear weap-

[a] With Joseph Rotblat.

ons were deployed for economic reasons. In 1954 NATO decided to deploy tactical nuclear weapons in order to compensate for its weakness vis-a-vis the Soviet bloc in conventional arms; to overcome this inferiority 96 divisions of conventional forces were thought to be required, and this would have been far too costly. [4] But later nuclear weapons became the cornerstone of NATO's strategy, the guarantor of peace. Having given credit to the deterrent action of nuclear weapons for the absence of war in Europe since 1945, the possession of these weapons became essential for the maintenance of peace. Various doctrines of that strategy, from mutual assured destruction to flexible response, were evolved, and a variety of ways of employing the nuclear deterrent were analyzed, but in all of them the retention of nuclear weapons was a *sine qua non.*

However, even the advocates of the doctrine of extended nuclear deterrence recognized the necessity of reducing the enormous nuclear arsenals and halting the arms race, either because of the fear of a pre-emptive first strike, or of accidental or inadvertent employment of nuclear weapons. Cuts in nuclear arms have been suggested down to a value that would be needed to prevent any type of war, by the threat of retaliation with nuclear weapons. For proponents of the "no-first-use" policy, a minimum deterrent is needed to guard against nuclear blackmail, to ensure against possible cheating. For them the value of the minimum deterrent is largely determined by the effectiveness of verification measures, and could be made vanishingly small if the verification regimes, discussed in the earlier chapters of this book, could be made sufficiently tight.

This paper looks at the longer perspective, after the implementation of the verification systems proposed in the earlier chapters. It argues that the permanent retention of nuclear weapons, even at a low level, is not acceptable for moral, political and strategic reasons. It then goes on to analyze ways to approach the goal of a nuclear-weapon-free world.

Nuclear Weapons and World Security

It is almost universally accepted that nuclear weapons are of no military value as an instrument for waging war. This is enshrined in the dictum: "A nuclear war cannot be won and must never be fought" agreed to by Presidents Gorbachev and Reagan during their first summit in November 1985. Nevertheless, these weapons are endowed with an important political and military role: they deter an adversary from using his nuclear weapons (simple deterrence) or indeed from using weapons of any kind (extended deterrence).

There is a vast literature on this subject and different approaches to it, but basically the arguments in favor of nuclear deterrence are predicated on the fact that there has been no war in Europe since 1945. It is alleged that this is due entirely to the existence of nuclear weapons and the threat of their use against an aggressor. Nuclear weapons are seen as a stabilizing factor, as "peacekeepers" (the name given to the U.S. MX missile). The invention of nuclear weapons has even been described as the most beneficial of all discoveries in the 20th century. [5] Furthermore, it is stated that since they cannot be disinvented, and the knowledge of how to make them will always be with us, there is no alternative but to retain them.

That peace has been preserved thanks to the nuclear deterrent is an unproven supposition. But continuous assertion in dogmatic terms (for example "Nuclear deterrence has certainly prevented world war—that world war which would otherwise have inevitably broken out sometime, somewhere, after 1945 between America and her allies and Russia and hers" [6]), repeated by others in parrot-like fashion, has given it the aura of unshaken credibility and acceptance as the gospel truth. Actually, it can be challenged on several grounds.

There is a moral argument against nuclear deterrence. Considering the enormous destructive power of nuclear weapons, exceeding by orders of magnitude any other type of weapon, the very concept of nuclear deterrence runs into moral difficulties. If the purpose of nuclear weapons is to deter their use, or the use of other weapons, by your adversary, the latter will take the threat seriously only if you are prepared to use nuclear weapons in retaliation, whatever the consequences. You must show by your posture that you are prepared to do this; otherwise your adversary will conclude that you are bluffing. There cannot be half-hearted deterrence. But does any cause exist that would morally justify the killing of tens or hundreds of millions of innocent people? How can one reconcile the natural repulsion of committing genocide with the potential undertaking to carry out such a deed? The moral argument is usually dismissed by the proponents of deterrence as being of no significance in military considerations. "How many divisions does the Pope have?" But this ignores the fact that the whole fabric of civilized society is based on moral values. If these are violated in one important area, how can they be defended in others? Security achieved by a balance of terror is bound—in the long run—to erode the ethical basis of our civilization.

There is also a political argument. The assertion that the absence of war in Europe since World War II is due to the nuclear deterrent ignores two facts: one, the many bloody wars outside Europe, in which some 22 million people have perished and in some of which nuclear-weapon states were directly involved; and two, the completely changed political configuration in Europe since that time. There is no evidence based on history that war in Europe would have occurred without nuclear weapons, or that war is bound to occur as soon as these weapons are removed. One can argue to the contrary, that the existence of nuclear weapons is responsible for the state of tension and mistrust that existed between East and West, and that this tension, and, in consequence, the probability of a non-nuclear war, would have been reduced if these weapons were removed. In any case, the concept of extended deterrence, that it prevents acts of aggression with non-nuclear arms, is basically weak since it rests on the implicit assumption that leaders will behave rationally and will not embark on ventures that may turn out to be suicidal. There is enough historical evidence that irrational people do come to power, even if not frequently.

Nuclear weapons cannot be disinvented, but this does not mean that they have to be deployed, or kept in storage, and that research must continue to invent "better" weapons. The same applies of course to chemical and biological weapons.

Has nuclear deterrence provided stability? The evidence from the history of the nuclear arms race points to the opposite. [7] At no time since the arms race began in 1949 has either of the superpowers been satisfied that what it had in its nuclear arsenals was sufficient to give it security. Each side felt compelled to keep on increasing and modernizing its arsenals, so that they have grown to the present monstrous sizes. Two main reasons for this are: technological advances and changing strategic concepts. Scientific and technological progress of research in military establishments keeps eroding the value of the deterrent, making it vulnerable and raising the specter of a first disabling strike; this creates the need for further research and for modernizing the weapons; a continuous arms race. Moreover, in the name of making the deterrent more credible, strategic concepts are changed, for example, from counter-city to counter-force, and this necessitates the introduction of new categories of weapons. But the strategy of flexible response does not make the nuclear deterrent any more credible. Even if only military installations were targeted, their proximity to population areas would result in enormous civilian casualties, particularly in Europe. A recent study has shown that even a very limited nuclear exchange in Europe, with less than 100 megatons on military targets, could result in more than 100 million casualties, mostly fatalities. [8]

The result of the unending arms race, in the quest for a reliable deterrent, was an increased probability of a nuclear war starting accidentally, by miscalculation or misperception. This is due to the ever increasing reliance on computers and other sophisticated technology, which may go wrong, and the greatly reduced time interval in which a decision has to be made about response to a perceived attack. Several events from the recent past have added to the serious worry about such responses in time of crisis.

The best evidence of the lack of reliance of nuclear deterrence to ensure security was President Reagan's SDI project, originally designed to protect populations by intercepting incoming missiles. [9] Even if it were technologically feasible—an assumption completely unwarranted according to the majority of independent expert opinions—it would destabilize the situation even further by encouraging the other side to augment its offensive arsenals, quite apart from the inherent high degree of instability of space weapons. The recent proposals from U.S. defense analysts of a new strategy, called "discriminate deterrence," is further evidence of the failure of the current deterrence policy. [10]

The Minimum Deterrent

The dangers inherent in the current strategic policies have been generally recognized, and even ardent advocates of nuclear deterrence have accepted the need to halt and reverse the arms race. Hence the efforts to reach agreements in international negotiations on the reduction of military arsenals of all kinds, nuclear, chemical, conventional, as well as on confidence-building measures. The first outcome of these efforts is the INF Treaty signed in 1987. [11] Negotiations have been going on since then on a treaty to reduce the strategic nuclear arms by 50%, in the first instance (the actual reduction is likely to be about 30%).

These developments are of great importance and were justly acclaimed by all who desire genuine nuclear disarmament. But the enthusiasm is tempered by concern that the gains may be illusive. There is a real danger of circumventing the INF Treaty, under the pretext of modernizing other categories of weapons; in effect, this might mean a return to the status quo ante even in the range of weapons to be eliminated. In addition, new weapons coming into deployment as a result of modernization programs of recent years will more than make up for the agreed reductions. In this connection it should also be recalled that the INF Treaty does not include the actual destruction of the warheads, and either side is permitted to transfer them to other missiles.

Nevertheless, a start has been made on nuclear disarmament. The important questions now are: how far should it go? what other types of weapon should be reduced, and by how much? what should be the ultimate aim of these efforts?

The advocates of the concept of extended deterrence would agree to a reduction of nuclear weapons to a value which, while posing a smaller risk of a first strike, would still retain the potential of adequate retaliation to an attack with nuclear or other types of weapons. This minimum deterrent would, therefore, have to contain enough weapons of different categories to satisfy the given strategic policy, e.g. flexible response. But whatever the contents, a certain number of nuclear weapons will have to stay in the arsenals as long as there exists a risk of military confrontation; in the opinion of some politicians this means in perpetuity.

For the proponents of simple deterrence, i.e. threat of retaliation against the use of nuclear weapons, a certain number of these weapons would have to be retained as an insurance against cheating. Such an insurance might be provided by one class of weapon, the most invulnerable one. The actual value of the minimum deterrent would be determined by the effectiveness of the verification system designed to prevent cheating and able to detect early enough any attempt to violate agreements. With improvements in performance of the verification system, the minimum deterrent could be gradually decreased.

In the early stages of the disarmament process, this approach would not differ in practice from that called for by the proponents of a nuclear-weapon-free world. They too would like to ensure compliance with agreements to eliminate nuclear weapons. But the difference in the basic philosophy about nuclear deterrence makes them seek the achievement of their objective by bolder measures.

A quantitative evaluation of the minimum deterrent policy has to include types, numbers and time of implementation. Concrete proposals about these have been made by Richard Garwin. [12] His main concern is the increasing instability of the current system associated with the counter-force strategy, and he sees deMIRVing as the key to deep cuts. By the single stroke of removing all but one warhead from present MIRVed launchers, the number of strategic warheads held by each side could be reduced from about 10,000 to 2,000; this could be accomplished in two years. Garwin recommends a minimum deterrent of 1,000 warheads each for NATO and WTO, the number to consist of 400 ICBMs with single warheads (the Midgetman in the USA), 400 SLBMs, carried on 50 submarines, and 200 air-launched cruise missiles carried on 100 aircraft. The whole process of disarmament to these levels could be achieved over a period of ten years in the following order of dismantling: all non-strategic missiles; reduction of MIRVs to one warhead per silo; elimination of half of the remaining launchers.

A lower minimum deterrent, about 600 warheads for each side, was suggested by the Committee of Soviet Scientists for Peace and against the Nuclear Threat, which carried out a detailed study of the problem. [13] They want to see the whole deterrent based on ICBMs, some of them mobile, with single warheads.

At a Pugwash Workshop in June, 1988, a still lower value for the minimum deterrent, to be achieved by the year 2000, was discussed, namely, 500 invulnerably based weapons by each side. For the NATO side the distribution of the warheads was envisaged as follows: 200 in the United States, 100 each in France and UK, and 100 non-strategic weapons to be held in Europe. [14] This last item implies retention of nuclear weapons for the purpose of extended deterrence.

If the above value of the minimum deterrent, a total of 1000 warheads, were achieved, it would represent a reduction to less than 2% of the present nuclear arsenals, a truly remarkable accomplishment. However, the remaining arsenals would still contain about 200 megatons of explosives, 25 times more than has been used in all previous wars. Even this very low "minimum" would have been considered unacceptably high 30 years ago when the notion of minimum deterrence was first discussed. It is certainly not acceptable as a permanent solution by those opposed to the whole concept of deterrence, and who are convinced that a finite—even if very low—nuclear deterrent is no more likely to be stable than the current nuclear arsenals.

One of the doubts about the stability of the minimum deterrent is whether we can really conceive that the deterrent will be maintained constantly at the agreed minimum value. We are contemplating going back to a situation, in terms of numbers, that existed many years ago, but is there reason to believe that this time the arsenals will not start growing as they did then? As long as either side has nuclear weapons, however small in number, there will be the need to maintain them. A core of scientists and engineers will be necessary for this purpose. And, as is their habit, they will think of ways of making their own weapons more effective, and those of the other side more vulnerable. A nuclear arms race is most likely to start all over again. It is very doubtful whether the minimum deterrent will stay at the minimum for long, unless there is an even more radical change in the political climate than has occurred recently.

Another question is, who will be entitled to possess the minimum deterrent? The schemes described above envisage the deterrent being retained by NATO and WTO, which effectively means the two superpowers. Some schemes allow the retention of some nuclear weapons by China, France and the UK. But what about the other countries, like Israel and South Africa, which may already have in their nuclear arsenals more than allocated to China, France and the UK, although undeclared? Will they be given tacit approval to retain these weapons, or will they be forced, by military, political, or economic means, to give them up? Neither of these solutions seems likely.

This raises a more general question. If the permanent retention of nuclear weapons is allowed for some countries, in order to make them immune from attack with any type of weapon, then how can one deny the same means of immunity to other countries? Such a denial would contain an inherent element of division, and therefore of mistrust and tension in the world: in our hands nuclear weapons are guarantors of peace, but their possession by other countries would increase the danger of war. According to article VI of the Non-Proliferation Treaty, the nuclear-weapon states undertook to pursue the aim of nuclear disarmament, as part of the ultimate objective of general and complete disarmament; there is no provision in the NPT for a permanent retention—even if at a very low level—of nuclear armaments. Horizontal proliferation of nuclear weapons has been kept under control so far, but this cannot go on for ever. What is likely to happen is that the NPT Treaty will be scrapped, all restraints it contains will be removed, and a number of countries will acquire the "minimum" nuclear deterrent. Some of these countries are sufficiently unstable to make us view this prospect with much anxiety.

It has been suggested that if the disarmament measures included a comprehensive test ban, this would prevent a build-up of nuclear arsenals by other countries. However, the testing of weapons is no longer a necessary element in the attainment of a nuclear arsenal. The technology of manufacturing bombs is now so well known, that they can be put in the arsenals as reliable weapons, without testing them. [15] Witness the case of Israel: everybody accepts that it has a credible nuclear arsenal even though Israeli tests have never been reported.

The retention of nuclear weapons—at a minimal level—is being justified on the basis of the premise that the existence of these weapons has secured peace in Europe and will reduce the risk of war in the future. Even though this premise is unproven, its plausibility is sufficiently high to make many peace-loving people accept nuclear deterrence as the lesser evil. But in this interdependent world, we must be concerned not only with avoidance of war in Europe but everywhere else. The extension of that premise to other parts of the world, where wars have been raging all the time, would justify the acquisition of nuclear arsenals by other states, unless the superpowers, plus perhaps the other three nuclear-weapon states, take upon themselves the role of a world police, intervening in every military conflict, a role far exceeding that of the Security Council. This is likely to create even more tension and generate violent reaction, and would in any case not be acceptable even to the citizens of the nuclear-weapon states. When balancing the possible risks of military confrontation resulting: (a) from such tension, (b) from the removal of the nuclear deterrent, the former seems to present the greater risk.

Elimination of Nuclear Weapons

Having arrived at the conclusion that the retention of nuclear weapons in perpetuity will not guarantee peace in the world—and may in fact undermine it—one needs to discuss the feasibility of the alternative policy, the elimination of nuclear weapons.

There are two dimensions to this problem: (a) technological, i.e. insurance against a violation of treaties involving complete nuclear disarmament; (b) political, i.e. build-up of the necessary climate of trust. Both are essential, and, as will be shown, interact and reinforce each other.

Verification of Reductions to Very Low Values

A frequent argument against complete nuclear disarmament is that even if all existing weapons were dismantled, it would not take long to reintroduce them, should this be required by a military or political contingency. However, even if the incentive to use nuclear weapons did arise in a nuclear-weapon-free world, the situation would still be much safer than if nuclear weapons were ready in the arsenals. If military establishments had stopped research on nuclear weapons (an essential element of nuclear disarmament), it would take some months for either side to manufacture and deploy the weapon, particularly if no weapon-grade plutonium or uranium were kept in storage, and this would provide time for negotiations to end the conflict by peaceful means.

The main difficulty about complete nuclear disarmament is the worry about cheating. The destructive potential of nuclear weapons is so enormous, that the concealment of even a small number of such weapons could give the side that has done this an overwhelming advantage. Therefore, we have to address the problem of verification of compliance with undertakings during the process of nuclear disarmament.

Obviously, getting rid of nuclear weapons will have to be a lengthy and gradual process, strictly correlated with verification effectiveness. The connection between reduction of nuclear arms and verification is usually described by the so-called Wiesner graph, first put forward at a Pugwash Conference. [16] A modified form of the Wiesner graph is shown in Fig. 1.

The vertical axis on the left is a measure of the status of armaments, for example, the number of nuclear weapons held by one side, or some other quantity characterizing the potential of these weapons. The horizontal scale gives the time from the start of the disarmament process. Both these scales are arbitrary. Let line A indicate the number of weapons held at a given time by our adversary, as declared by him. Natural caution will make us assume that he is cheating, that he managed to conceal some weapons. Let line B be the real number held by him. Note that at the beginning, when the arsenals are still large, even a considerable concealment would not matter, but the situation changes as the arsenals become smaller. Even a small amount of cheating could then be decisive. Therefore, the necessary condition to proceed to low levels of armaments is effective verification of compliance. The state of necessary verification is indicated by line C, with its percentage scale on the vertical axis on the right. The shaded area between A and B represents the degree of uncertainty at a given stage of disarmament; it is inversely related to the effectiveness of the verification procedures achieved at that time.

It should be noted that Fig. 1 is an idealized presentation. In reality, neither the process of nuclear disarmament nor the increase in the effectiveness of the verification system would be linear functions of time, but it is the general trend that matters.

The main difficulty arises when the level of declared armaments approaches zero. Either side will agree to take the last step of complete disarmament only after the effectiveness of verification has reached 100%. But since this can never be assured, a different approach will have to be adopted when that stage is reached. This is discussed below.

Elimination of Missiles

There are several ways to achieve arms reduction. One of them is reduction across the board, i.e. all types of weapons are reduced by agreed amounts in any stage of the disarmament program. This type of approach is being attempted, in relation to strategic weapons, in the "50%" cut negotiations in

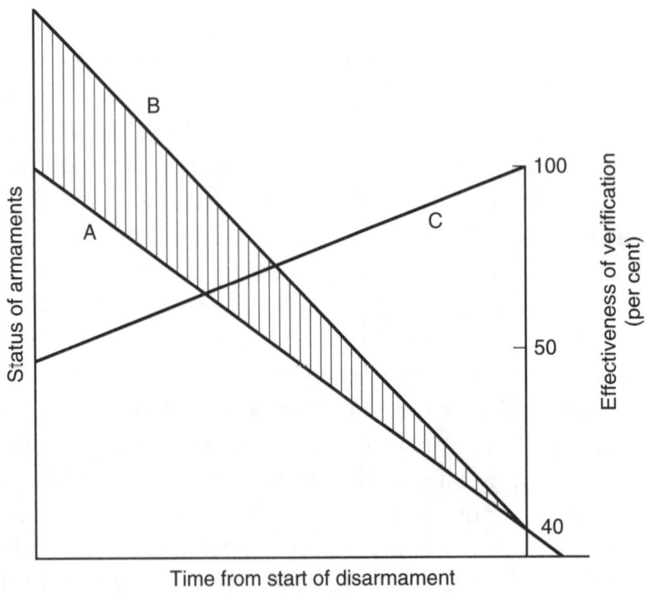

FIGURE 1.

START, and has encountered many difficulties. A more efficient way would be the complete elimination of weapons in a given category, one at a time, as was shown by the relatively quick agreement on the INF Treaty, which is characterized by thoroughness and precision. Obviously, it is easier to verify the complete removal of a given class of missiles than to reduce them gradually by agreed numbers or proportions.

Bearing in mind the probable requirement to maintain an invulnerable deterrent during the process of disarmament, and the anxiety not to lower the nuclear threshold in the meantime, the following sequence of eliminating weapons seems indicated. First, the short-range missiles, in their great variety as numbers of theatre or tactical nuclear weapons, should be destroyed. In the completely changed political map of Europe their deployment has become an anachronism. Indeed, during the transition period, with its inherent instability, the possession of such weapons by troops is potentially highly dangerous. A reduction and restructuring of conventional forces, preceding or simultaneous with that step, would of course be of great value in this respect.

The next weapons to be eliminated would be land-based ICBMs, followed by missiles and bombs carried by aircraft, and lastly submarine-launched missiles.

The specific techniques required to verify that the missiles and warheads have been destroyed, such as sealing, tagging, and "fingerprinting", are described in Chapter 5 of this book.[1]

Disposal of Nuclear-Weapon Materials

Unlike the INF Treaty, the program of complete nuclear disarmament calls for the elimination of warheads, as well as missiles. In a nuclear-weapon-free world there is no need for nuclear warheads. The main problem is the disposal of the special materials needed to make nuclear weapons, i.e. highly enriched uranium, plutonium, and tritium.

A prerequisite for a program of nuclear disarmament is a halt to the production of these materials. The requirements for verifying a production cut-off of such materials are presented in Chapter 6.[2]

The amounts already in existence are estimated to be 1000 tons of 90 percent uranium-235, 200 tons of plutonium, and 200 kg of tritium, these quantities being about equally shared between the USA and USSR. [17] The relatively short half-life of tritium (12.3 years) ensures its natural elimination

[1]Chapter 5 not represented here.

[2]Chapter 6 not represented here.

within a few decades (down to one-fifth in 30 years). With regard to the other two materials, there is the legitimate desire to make use of their very high energy content while they are being rendered unsuitable for military purposes. [18]

The problem of uranium is simple. Its dilution with uranium-238, so as to reduce the U-235 content to about 4%, ensures that it is no longer a weapon material, but highly suitable as fuel in thermal reactors. The disposal of plutonium is more difficult, because there is no practical way of rendering it militarily useless by isotopic dilution. The suggested method of getting rid of the plutonium without wasting its energy content is to fabricate fuel consisting of mixed oxides of uranium and plutonium for existing thermal reactors.

The burning of plutonium in such reactors is accompanied by the production of new plutonium from uranium-238; the overall rate of eliminating plutonium stores would therefore be rather slow. Since the use of mixed fuels would in any case require a change in design of reactors due to problems of reactivity control (see Chapter 5 of this book[3]), it may be sensible to go a step further and build reactors which use plutonium as the only fuel, i.e. without uranium. This will be more costly, but the expense is justified in view of the very large stockpile of plutonium. The advantageous feature of such reactors is that they would be the only generators of nuclear energy (until thermonuclear reactors come into being) which do not at the same time produce new plutonium. The plutonium remaining in the spent fuel elements would be intrinsically mixed with the fission products and thus inaccessible for a very long time. The fabrication of the plutonium fuel for such reactors would have to be carried out under very exacting safeguards, and the control of material flow would have to be very strict, to prevent loss of plutonium by theft or by terrorist acts.

Another way of disposing of the plutonium in the arsenals is to recycle it in molten-salt reactors, in which the fuel (in liquid state) would consist of a mixture of thorium and plutonium fluorides. [19]

Apart from the plutonium in nuclear warheads, there are huge quantities of plutonium produced in electricity-generating stations based on thermal reactors. Although the Pu-240 content of such plutonium is too high to classify it as weapon-grade material, it could still be used in nuclear weapons. [20] For this reason it would be desirable that the existing stores of plutonium from civil reactors, separated in reprocessing plants, should also be burned in the special reactors mentioned above.

As an independent measure to prevent further availability of plutonium, the practice of reprocessing spent fuel elements from reactors should be aban-

[3]Chapter 5 not represented in this book.

doned in favor of the "once-through" cycle, i.e. storage without processing. The highly radioactive fission products in the spent fuel constitute an effective barrier to access to the plutonium.

The measures proposed here would not adversely affect the economy of nuclear energy production, since the era of the fast breeder reactor—for which separation of plutonium is needed—is receding further into the future, mainly due to the glut of uranium in the world. The extra uranium becoming available from the military stockpiles would add to this glut.

Dual-Purpose Weapons

If the materials for nuclear weapons were eliminated, the existence of missiles that could be converted to carry nuclear warheads would be much less of a problem. Nevertheless, methods of verifying that dual-purpose missiles are not equipped with nuclear warheads would have to be included in a program of nuclear disarmament. Again, the simplest solution is to eliminate such weapons altogether, in a complementary program of conventional disarmament.

Difficulties with the Complete Elimination of Nuclear Weapons

The main concern about establishing a nuclear-weapon-free world is how to ensure that one side will not be cheating in the last stages of nuclear disarmament, and conceal a small number of such weapons with which it could gain political power, if not world domination. The verification techniques, discussed in this book, could ensure compliance with agreements to reduce weapons to very low levels, but cannot be 100% effective. Although full effectiveness is generally not required, the enormous destructive power of nuclear weapons makes this a necessary requirement when complete nuclear disarmament is reached.

One of the major causes of uncertainty is the amount of plutonium that might be concealed. Despite the stringent accountability of materials in nuclear facilities under IAEA safeguards, a certain amount of plutonium is lost in the various stages of the nuclear fuel cycle; this comes in the category of MUF (material unaccounted for). It would be difficult to reduce MUF to much below a few percent, and this may add up to a few tons of plutonium. Only a fraction of this might possibly be concealed by any one state, but this would be sufficient to make a number of bombs, theoretically enough to blackmail another state into submission.

The most radical way to deal with this is to remove the incentive for such an act by creating a climate of relations between nations such that nobody will want to threaten others. But such a climate, described by one scientist as "nuclear weapons: who cares?", [21] will never be reached as long as states possess nuclear weapons. As argued earlier, if nuclear weapons were retained by the present five states, or even by the two superpowers only, this would be a constant irritant, a perpetual cause of tension that would preclude the establishment of the necessary political climate. If many states were allowed to have nuclear weapons, the situation would be even more precarious, because there can be no guarantee that an irrational person will never become the leader of a state. Another possibility is the acquisition of nuclear weapons by anarchic groups in a civil war which might follow, for example, the collapse of prerestroika in the Soviet Union. [22] The notion that we could live with nuclear weapons and nobody be concerned about them is absurd; the risk involved is far too great to ever be acceptable. For this reason we may have to turn to a different measure to ensure that no state will resort to threatening the use of nuclear weapons acquired or made clandestinely.

Such a measure would be the setting up of an international authority endowed with a small nuclear weapons capability. In a way, this resembles the Baruch Plan to create an International Atomic Development Authority. [23] But the world has changed since 1946, when anything even remotely resembling a World Government was rejected out of hand. We are still not ready for World Government—although there is much less adversion to it—but there is now much more respect and trust in joint international undertakings. The United Nations has proved itself in a variety of policing duties and peace negotiations, and has acquired enough status to be entrusted with an even mightier task, to be the custodian of an instrument designated to prevent a threat to world security.

Specifically, the proposal is to create a new organ of the United Nations, to be the keeper of a small number of nuclear weapons and means of delivery, with an international police force to make the cache invulnerable. The purpose of this UN nuclear arsenal would be to prevent a state which suddenly disclosed the possession of a nuclear armory from threatening its use as a political instrument of submission of another state. To be effective, the decision of the UN organ to initiate retaliation must not be allowed to be overridden by the veto of one state, which might itself be the offender or its sponsor. A decision by n-x of its n members, where x might be 1,2 or even higher, depending on the value of n, would be sufficient.

An intermediate step, suggested by Garwin, would be a formal agreement by nuclear-weapon states that weapons possessed by them would be used in retaliation only *through* the United Nations. [24]

Inasmuch as the scheme outlined above relies on deterrence, it is abhorrent to those who object to the very concept of deterrence on moral grounds;

it could not therefore be accepted as a permanent solution. As a transitional measure it has the appeal that it would reduce the number of nuclear arsenals from five to only one. It should also be noted that the only purpose of the single arsenal would be to prevent a nuclear threat, and that it would come into being at a time when nations have pledged themselves to get rid of their nuclear weapons. The need for the international depository of nuclear weapons will diminish as time goes on, because of better technological means of verification being developed and a conducive political climate.

The Political Climate

The discussion of the ways toward a nuclear-weapon-free world has so far been based on the technological aspect of the problem. But the political aspects are just as important. Indeed, in the opinion of many, the whole problem is a political one; the decision to use, or to threaten to use, nuclear weapons is said to be entirely political, and that therefore the solution must be sought in political terms.

While it is true that the decision whether to employ a weapon is political, such a decision would be meaningless unless the state had the technical wherewithal to implement it. Similarly, the political decision to eliminate certain weapons by a treaty would lack credibility unless there were the technical means to verify compliance with the treaty. [25] Both aspects, the technological and the political one, are necessary elements. Moreover, not only do they both play a vital role, but they reinforce each other.

The pursuit of an arms race, and the existence of large military arsenals, go hand in hand with a climate of mistrust, antagonism, and fear. One is the father and the child of the other. In order to justify the enormous military spending, it is necessary to create the image of an enemy, who is depicted as evil and aggressive, waiting for any sign of slackness in our defense posture to pounce and overwhelm us.

Conversely, the mutual reduction of arsenals alleviates anxiety and creates trust and good will, which is conducive to further measures in the same direction. It is a process of positive feedback. Disarmament measures reduce fears and bolster confidence, which in turn makes it easier to implement further reductions, which create a better climate, and so it goes on.

In terms of the Wiesner graph (Fig. 1), which was based only on the technological aspect, the inclusion of the effect of an improved political climate would accelerate the process of disarmament, by changing the idealized straight line A (Fig. 2) into a downward bending curve A'. Again, these graphs are meant to show trends and do not imply meaningful mathematical relations.

The above considerations apply not only to nuclear weapons but to all

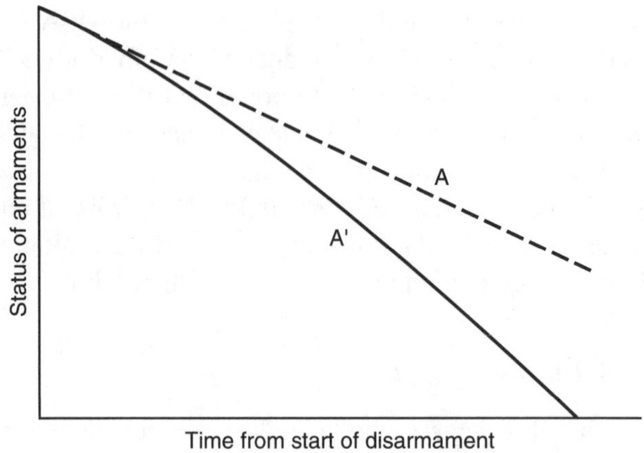

FIGURE 2.

types of military arms. Indeed, it would be folly to aim at nuclear disarmament in a world increasingly armed with other weapons. Nuclear weapons have been singled out because of their enormous destructive potential; should these weapons be used they would threaten the whole fabric of our civilization and the lives of most of the world population. The removal of this menace must therefore have top priority. But the process of nuclear disarmament must be accompanied by steps toward the reduction and eventual elimination of stockpiles of chemical and conventional weapons. Recent moves in this direction, for example, the unilateral decisions by the Soviet government to reduce significantly the numerical strength of its army and the numbers of its conventional armaments, [26] as well as its stock of chemical weapons, [27] and the statement by President Bush in the same direction, [28] are highly encouraging. Here again the positive feedback between technical measures and better international relations would greatly help in the overall eradication of the threat of military confrontation, eventually leading to the ultimate objective, general and complete disarmament. [29]

The Role of Scientists

The scientific community has an important role to play in helping to devise and improve the ways to nuclear disarmament, as well as encouraging the creation of the necessary political climate.

It is appropriate to recall Einstein's statement of 1946 soon after the birth of the nuclear age: "The unleashed power of the atom has changed everything save our modes of thinking, and thus we drift toward unparalleled catastrophe. We scientists who unleashed this immense power have an overwhelming responsibility in this world life-and-death struggle to harness the atom for the benefit of mankind and not for humanity's destruction." [30] With the recent evidence of a change in the mode of thinking on these issues by political leaders, it is even more incumbent on scientists to actively fulfill their social responsibility.

In the first instance, scientists working in military-research establishments can help directly. If nuclear and other arms are to be eliminated, the existing establishments will have to cease working to prevent a renewal of the arms race. But they could be kept going if they were converted entirely for peace research, specifically, to enhance verification techniques. Some research of this kind is already going on there, in parallel with work on new weapons, but it would have to be expanded considerably. The scientists and engineers should be asked to use their talents to discover new methods of surveillance, new techniques to identify missiles and warheads, new ways to reduce the likelihood of undetected cheating.

Other scientists can also help in these tasks. The contributors to this book—nearly all independent scientists not connected with military establishments—have already shown their ingenuity in advancing verification techniques. Much more could be achieved if a larger number of scientists put their minds to this problem. [31]

Another role of the scientific community is to act as a watchdog of compliance with treaties. Through their contacts with other scientists in universities and research institutes, and through careful reading of the literature, they can discover whether clandestine work is being carried out, or sensitive material is being diverted, in this way lessening the possibility of cheating. A network of information about suspicious events, including a register of scientists and technologists, will strengthen societal means of verification. And the very knowledge that scientists are on the alert would help in reducing anxiety among the public about non-adherence or circumvention of treaties.

In this context, there is also a significant role for scientists in educating the public about the importance of verification of treaties, and preventing their violation, as part of common security. Several specific measures toward this objective have been proposed. [32] There is a need to foster a general feeling of loyalty to the world community as an extension of the loyalties to one's country and family. Due to the universality of science, which cuts across geographic and ideological boundaries, scientists have developed this sense of belonging to the world community, of being citizens of the world, more than any other group of society, and therefore they are in a good position to nurture such ideas.

Another way for scientists to contribute to the formation of a suitable political climate is by promoting and taking part in global projects of scientific collaboration. Such projects, involving scientists from many countries, not only serve to improve the image of science, but the result of such collaboration can be of great benefit to the peoples of the world, raise standards of living, and create more goodwill on earth, thus lessening the need to resort to military solutions.

Conclusions

Two questions were posed in the title of this paper: is the elimination of nuclear arsenals desirable; and, if so, is it feasible?

The answer to the first question is an emphatic yes. Quite apart from the moral objections to reliance on retaliation with weapons of mass destruction, nuclear deterrence is not a reliable means to ensure security, if based on current arsenals and the inherent danger of an accidental war. Nuclear disarmament down to a "minimum deterrence" will still not bring stability; in the long run a new arms race is bound to occur, and more nations are likely to acquire nuclear weapons for their security, thus increasing the risk of nuclear war.

The answer to the second question is a qualified yes. The known means of verification can ensure compliance with undertakings to reduce nuclear arsenals down to a very low level. With the present state of technology, they cannot be made 100% effective, a necessary condition for the complete elimination of nuclear weapons. Further research is needed to elaborate the approach to that state. As an interim measure, the treaty abolishing national nuclear arsenals would need to be accompanied by the setting up, under the aegis of the United Nations, of an international depository of nuclear weapons and means of delivery sufficient to deter a potential violator of the treaty.

Apart from continuing efforts to perfect the technological means of verification, the attainment of the objective of a nuclear-weapon-free world would be accelerated by the creation of a climate of goodwill and trust, itself a result of arms reduction. Scientists have a vital role in these tasks as well.

Notes

1. UN General Assembly, Resolution 1(I), 24 January, 1946.
2. Statement by Mikhail Gorbachev, General Secretary of the CPSU Central Committee, 15 January, 1986, Novosti Press Agency.
3. "Action plan for ushering in a nuclear-weapon-free and non-violent world order." Tabled by India at the Third Special Session on Disarmament, June 1988.
4. Lord Zuckerman, "The World Without INF." *The New York Review of Books*, 2 June, 1988, p. 37.
5. British Prime Minister Margaret Thatcher, in a B.B.C. interview, 7 December, 1987.
6. William Waldegrave, British Foreign Office Minister, *The Times*, 6 January, 1989.
7. J. Rotblat, "Technology, the Arms Race and Disarmament" in C. Schaerf, B. H. Reid &

David Carlton, eds., *New Technologies and the Arms Race* (London: Macmillan, 1989), pp. 332–344.
8. Andrea Ottolenghi, "Limited Nuclear War in Europe" in *Effects of Nuclear War on Health and Health Services.* (Geneva: World Health Organization, 1987), pp. 121–125.
9. Address to the Nation by President Ronald Reagan, "Peace and National Security," 23 March, 1983.
10. F. C. Iklé & A. Wohltsetter, "Discriminate Deterrence." Report of the Commission on Integrated Long-Term Strategy. January 1988.
11. "Treaty between the United States of America and the Union of Soviet Socialist Republics on the elimination of their intermediate-range and shorter-range missiles," December 1987.
12. R. Garwin, "A blueprint for radical weapons cuts," *Bulletin of the Atomic Scientists*, March 1988. See also *Proceedings of the 37th Pugwash Conference*, Gmunden, September 1987, pp. 267–280.
13. A. A. Kokoshin, "A Soviet view on radical weapons cuts," *Bulletin of the Atomic Scientists*, March 1988.
14. Report on 16th Pugwash Workshop on Nuclear Forces. *Pugwash Newsletter*, Vol 26, July 1988, p. 14.
15. V. I. Gol'danskiĭ in *Nuclear Tests: Prohibition or Limitation?* (V. Goldblat and D. Cox, eds.), Oxford University Press, 1988, pp. 332–334.
16. *Proceedings of the 6th Pugwash Conference*, Moscow, 1960, p. 298.
17. F. von Hippel, D. H. Albright, and B. G. Levi, "Quantities of Fissile Materials in U.S. and Soviet Nuclear Weapons Arsenals." *PU/CEES Report* No. 168, Princeton University, July 1986.
18. E. Amaldi, U. Farinelli, and C. Silvi, "The Utilization for Civilian Purposes of the weapon-grade nuclear materials that may become available as a consequence of nuclear disarmament." Workshop on International Security and Disarmament, the Role of Scientific Academies, Rome, 23–25 June 1988.
19. A. Lecocq, "Disposal of Fissile Materials and Verification" in J. Hassard, T. Kibble, and P. Lewis, eds., *Ways Out of the Arms Race* (Singapore: World Scientific Publishing Co., 1989), pp. 77–83.
20. T. B. Taylor, private communication.
21. V. Weisskopf, "A Change of Emphasis." *Proceedings of the 37th Pugwash Conference*, Gmunden, 1987, pp. 85–88.
22. V. I. Gol'danskiĭ, *Physics Today*, March 1990, pp. 53–55.
23. "International Control of Atomic Energy: Growth of a Policy." U.S. Department of State, June 1946.
24. R. L. Garwin, private communication.
25. V. I. Gol'danskiĭ and M. Intriligator, "US-Soviet Relations: From Confrontation to Cooperation Through Verificational Deterrence" in *Proceedings of the 38th Pugwash Conference*, Dagomys, 1988, pp. 196–202.
26. Mikhail S. Gorbachev, United Nations Address, 7 December, 1988.
27. Eduard A. Shevardnadze, Speech at the Paris Conference on Chemical Weapons, 8 January, 1989.
28. George Bush, speech at NATO meeting, 30 May, 1989.
29. J. Rotblat, "Time to Think Again About General and Complete Disarmament" in *Ways Out of Arms Race*, op. cit., pp. 195–200.
30. O. Nathan & H. Nordon, *Einstein on Peace* (New York: Simon & Schuster, 1980), p. 376.

31. J. Rotblat, "Dilemmas for Scientists with a Social Conscience" in *Proceedings of the 38th Pugwash Conference*, Dagomys, 1988, pp. 105–112.
32. T. B. Taylor, "Global Abolition of Nuclear Weapons—Verification of Compliance and Deterrents to Violation," in *Proceedings of the 40th Pugwash Conference*, Egham, U.K., 1990, pp. 257–270.

An International Nuclear Security Force[a]

The choice between a world with nuclear weapons and a world with none may appear at first sight a clear, black-or-white choice: either nuclear weapons exist, or they are totally eliminated. However, particularly in the later stages of reductions in the number of nuclear warheads, certain "grey areas" emerge. The purpose of this chapter is to discuss some of these grey areas. There is the question of an international nuclear security force to ease the transition from a world with national nuclear forces to a world in which there are no nuclear weapons at all. There is also the question of various forms of civil research which might still need some low-yield nuclear explosions, and the possibility that there could be some other peaceful uses for such explosions. In particular, there is a small but finite danger that the Earth might be struck by an asteroid, and the only method of preventing a major disaster might be the projection of a high-yield thermonuclear device into the path of the incoming body.

One initial question concerns timing: how long it would be before a Nuclear-Weapon-Free-World (NWFW) could become possible. It would be wrong simply to extrapolate the rate of reduction which is implied by the recent statements of the Presidents of the United States and Russia. Once the two powers have reduced the number of their warheads to a figure of around 2000–3000 each, further reductions will become more difficult, since the other nuclear-weapon states will have to be brought into the negotiations. In any case, the reductions presently envisaged will need at least 10–15 years for their implementation.

There is another time-related aspect: to what extent can nuclear weapons be totally eliminated as long as civil power plants are producing large quan-

[a] With Stanislav Rodionov.

tities of plutonium as by-products of the nuclear power industry? [1] It is very much more difficult to construct adequate safeguards against illicit production of nuclear weapons if production of plutonium or enriched uranium for peaceful purposes is permitted, while its use for nuclear warheads is forbidden.

Nuclear reactors produce about 20 per cent of the world's electrical energy. Various proposals have been made for designing nuclear power systems that are much less prone to diversion of nuclear materials for weapons than present nuclear plants and their supporting fuel-cycle operations. None of them avoid the production or use of heavy isotopes that would be much easier to extract from fresh or spent fuel than it would be to enrich natural or low enriched uranium sufficiently to become weapons-grade.

A long-term alternative to nuclear energy could be controlled thermonuclear fusion as a new and more powerful source of ecologically clean energy (without radioactive fission products). The recent success of British scientists with a Tokomak experimental device is a rather encouraging indication that controlled fusion might possibly be achieved within 30–50 years. This could be used as a rough estimate of possible timing for complete nuclear disarmament.

Another relevant period of time is the connection with the average lifetime of new nuclear warheads, which reportedly is about 20–30 years (for thermonuclear warheads that could be explained by the lifetime of the tritium used in the warhead).

Therefore, there can be no question of a transition to an NWFW during the remaining years of the 20th century. Whether we like it or not, we will be obliged to reconcile ourselves to the existence of nuclear weapons well into the first half of the 21st century.

The process of nuclear weapon elimination should begin now by banning the production for nuclear weapons of highly enriched uranium, plutonium, and tritium; by large reductions in the number of warheads; and by developing technologies for the treatment of fissile materials. It seems reasonable to assume that, down to a certain total of nuclear warheads, five states (the USA, Russia, the UK, France and China) will continue to have national nuclear forces. In addition, there are the covert possessors of nuclear weapons—for example, Israel, India, Pakistan, and some others, possibly including a few of the former Soviet republics. There is a risk, which is not negligible, that nuclear weapons might fall into the hands of criminal groups, for example, under the conditions which now exist in some parts of the former Soviet Union. The smallest models of modern tactical nuclear weapons weigh only approximately 50 kg and have a diameter of less than 20 cm—so they can be considered portable devices—and explode with a yield equivalent to approximately 1000 tons of TNT.

Safety

This raises the question whether a nuclear weapon could be detonated without the involvement of those who know the safety and code systems. For example, U.S. nuclear warheads are designed so that:

> *in the event of a detonation initiated at any one point in the high explosive systems (rather than multiple points, which would occur providing the authorization codes were properly entered and the environmental sensors registered the design launch-to-target sequence), the probability of a nuclear yield greater than 1.8 kilogrammes of TNT equivalent shall not exceed one in a million.* [2]

However, there is another danger: that a "one-point" explosion, which would result in the dispersal of very tiny particles of highly toxic, radioactive plutonium into the atmosphere. This problem has been investigated quantitatively. [2] It has been estimated that a one-point explosion would convert about 30 per cent of the plutonium contained in the warheads into a PuO_2 aerosol of respirable size. If we assume an accident involving a single warhead, and that this warhead contains about three kilograms of plutonium, then the resulting aerosol would contain around one kilogram of PuO_2.

The warheads probably contain 20–50 kg of high explosive (HE). Since normal HE is nearly twice as powerful as TNT, an accident could result in an explosion equivalent to 40–100 kg of TNT. The corresponding initial cloud sizes are: cloud-top height: 200–300 meters; cloud radius: 20–30 meters. The main risk from exposure to plutonium aerosol is from inhalation; the external radiation from the passing cloud and from plutonium deposited on the ground can be neglected. Health effects from radiation can be divided into two categories: illness and death due to high doses, occurring within a month or so after exposure, and cancers occurring during the remainder of the lives of the exposed population, starting a few years after exposure to low doses. High-dose effects would result from exposure within about a hundred meters from the explosion point. The estimate of low-dose effects suggests that under worst-case weather conditions and for high population density, an accident with a single plutonium warhead—that is, detonation of HE with no nuclear yield—could result in a few thousand fatalities during the remainder of the lives of the exposed population. This is just the effect of the passage of a cloud of plutonium aerosol. To avoid more casualties, there would have to be evacuation procedures and cleanup of the area.

However, inside a pressurized chamber a one-point explosion of HE in any warhead can be used to obtain a critical mass of any fissile material after some chemical manipulation. In this way, any state—or indeed a terrorist group (with the help of appropriate specialists)—could convert normal safe warheads into nuclear feedstock for the construction of a new nuclear weapon, which might be more primitive but which would still produce an explosion.

UN Nuclear Security Forces

Once nuclear warhead numbers have been substantially reduced, the next step should be a qualitative one. Instead of national nuclear forces, there should be international nuclear safeguard forces in order to keep some stability during the transition to an NWFW. It should not be particularly difficult for the five nuclear powers to put their nuclear forces under international control. However, there could be a number of problems with the members of the "international underground." For example, Israel and some Arab countries might refuse to reveal their nuclear secrets unless the political crisis in the Middle East is in some way solved. There could be the same problems with India-Pakistan relations, or in the Korean peninsula. These examples show that progress in nuclear disarmament is highly dependent on the general world political situation. In many parts of the world an improved political climate is a necessary condition for the process of reduction of nuclear forces.

The idea of an international nuclear police force under UN auspices has been a matter of discussion for a number of years. [3] The proposal was to create a new UN organ, which would keep a small number of nuclear weapons and means of delivery. This would seem to be an inevitable stage in the nuclear disarmament process, for the Security Council and the Military Staff Committee would have the capabilities for the immediate and effective use of this force against any aggressor possessing nuclear weapons or any other weapon of mass destruction. These international nuclear forces could be called the UN Nuclear Security Force (NSF). This force would be created against the background of massive reductions in the nuclear weapon stockpiles, the guaranteed and controlled total destruction of chemical and biological weapons, considerable reductions in conventional arms, progress in creating effective European and Asian-Pacific security structures, and general and effective international means of conflict prevention and resolution.

Even under these favorable conditions it might be possible for some criminal groups or even governments to use clandestine stocks of nuclear materials, or new low-scale devices for isotope separation (which could not easily be verified by international inspection) for the construction of a few nuclear explosive devices; these might then be used for blackmail. Similar attempts could be made to produce other weapons of mass destruction. What should be the role of the NSF in these cases? Should its role simply be that of deterrence? The question then arises whether in all cases deterrence would be effective.

History suggests that deterrence works in relations between developed countries, where the adversaries have a rather large number of nuclear warheads and a developed political and economic superstructure. However, there are historical examples with developing countries where deterrence was ineffective, for example the Korean War and, in part, the Gulf War. Deterrence would not work against fanatical religious groups, or possibly even against some developing countries with low population density. So there could be cases where the NSF had to use nuclear weapons for a preventive or retaliatory strike. Of course, this should be the last resort for dealing with a threat to world security. There should first be political and economic pressure, and probably an attempt to deal with the matter by the use of conventional weapons.

International problems arising from an NWFW Treaty should not in general call for immediate action by the UN Nuclear Security Force. There should be lengthy attempts to manage such situations without resort to the use of nuclear force. Thus, the delivery systems which might be more relevant to this "slow response" policy would be either air-launched or sea-launched cruise missiles. Both are highly accurate weapons, and with further development their accuracy could be improved still more.

It should not be necessary for the NSF to conduct any nuclear tests (though later in this chapter it is suggested that there may be scientific reasons for permitting some low-yield tests). For purposes of weapons production, "nuclear weapons that derive all their energy from fission can now be developed with high confidence, without any nuclear tests." [4]

A "slow response" strategy would make it possible to take a succession of steps which could serve as a warning to violators: a declaration that the "first-use" of a NSF nuclear weapon might be necessary; the installation of a cruise missile on a UN aircraft or ship; the movement of the delivery vehicle to a base which was within cruise missile range of a potential target, and so on. The arrival of a nuclear armed ship or aircraft should result in additional psychological pressure on the violator.

Any actual use of nuclear weapons by the NSF should be structured to specific ground-based or underground targets; a high degree of targeting precision would be needed. It could well be that in time non-nuclear explosives could be substituted for nuclear ones, as the risk of any violation of an NWFW Treaty decreased. A situation could be reached in which all highly enriched uranium and plutonium which had been produced before nuclear power was phased out had been rendered incapable of sustaining an explosive chain reaction, and in which detailed inspection gave a hundred per cent guarantee that there were no hidden stockpiles of fissile material. In that situation it should be possible for the NSF to dismantle the last nuclear warheads.

Nuclear Weapon Scientists and Research

Nuclear weapons cannot disappear without leaving any trace. One of these traces would be the remaining number of nuclear weapon specialists. This problem is already here; some countries which are seeking to develop nuclear weapons may attempt to recruit nuclear scientists from the former Soviet Union. The best solution is to avoid this potential brain drain by ensuring full employment for these specialists in international facilities under international control.

The total elimination of all nuclear warheads could not be followed by the immediate elimination of highly enriched uranium, since a number of civil technologies use this fissile material, for example, as fuel for nuclear submarine reactors or fuel for power modules of deep-space missions. All such activities should be under strict international control. There could, for example, be an International Fissile Material Center (IFMC), which would be responsible for the production of uranium fuel and for the construction of certain nuclear explosives.

The IFMC should probably be permitted to conduct, for peaceful purposes, nuclear explosions exceeding the yield level which has been suggested for a comprehensive test ban, that is, 100 kg of TNT. One activity which exceeds this threshold is inertial confinement fusion (ICF) in which microexplosions yield up to 300 kg of TNT, and could be contained in aboveground reactor vessels. [5] ICF could provide an experimental base for extensive studies of material effects and could be a key factor in maintaining a team of experts in fusion physics and diagnostics. Moreover, ICF microexplosions might also act as a useful research tool in the study of X-Ray laser physics (XRL).

The practical cut-off for explosions that can be contained in seismically quiet vessels is equivalent to some 300 tons of TNT, and indeed a design for a so-called High Energy Density Facility (HEDF) has been under study since 1981 by scientists at the National Livermore Laboratory in the U.S.A. This facility is supposed to be located at the Nevada Test Site, and would consist of a seismically quiet underground chamber able to contain nuclear detonations up to 0.3 kilotons. It would be reusable, operating with about one explosion a week. [6]

ICF studies on the design of imploding deuterium-tritium "pellets" and fusion ignition in these implosions may result in the design and construction of more powerful thermonuclear explosives which would not contain fission products and could be used for non-military purposes without the danger of radioactive pollution.

XRL studies together with gamma-ray laser (GRL) studies can provide a unique technique for producing genuine holograms of single atoms and molecules. At the moment it is not known what level of explosive yield is needed in order to get a genuine understanding of the laser generation process: it could be around 1 kiloton.

There are a number of proposals for peaceful nuclear explosions in addition to the XRL and GRL studies. There is the idea of creating large cavities for the safe containment of sewage and ecologically dangerous wastes. [7] A number of studies suggest that within 40–50 years practically all important petroleum sources will have been exhausted. Underground nuclear explosions could be a method of improving the productivity of old oil wells.

The conclusion here is that low-yield tests have a number of important non-military scientific applications, and that peaceful underground nuclear explosions may also be useful in some applications. Here a proposal made by one of the authors (VG) is relevant [3]:

> *As for underground tests, if there is no way the United States can renounce them, let us reach agreement on conducting all such tests jointly, gradually reducing their frequency and the upper limit for the permitted yield of the explosion. Where? In Nevada, for example, since the authorities of that state gave their consent to the functioning of a nuclear test site there. I believe that in this way we could really help to strengthen mutual trust, bring substantially nearer the complete cessation of nuclear tests, strengthen the Non-Proliferation Treaty, and make potential aggressors listen to reason: in other words, create the necessary preconditions for transition to a nuclear-free world.*

A zero-nuclear-weapon level in an NWFW is not, per se, a self-stabilizing system, since any weapons-related fluctuations can introduce serious political and military disturbances. General physical laws suggest that, for stability, a system needs strong negative feedback. After the need for a UN Nuclear Security Force has disappeared, the IFMC could provide negative feedback of this kind. Its international team of scientists could, if necessary, be in a position, at short notice, to produce nuclear weapons again, so that there would exist not real but "virtual" nuclear weapons which could be developed if the need arose.

One possible need is for the protection of humankind from the catastrophic collision of a huge meteorite, or comet nucleus, with Earth. The probability of such a collision for a Tunguska-type body (mass of the order of 10^5–10^6 tons, size of the order of a few hundred meters) is considered a low one, but it still has a finite value. Quite recently (March 1989) an asteroid 1989 fc was discovered to have passed within one million kilometers from the Earth. If it had collided with our planet, the effect would have been equivalent to a 5-megaton nuclear explosion.

The only realistic way of protecting humankind from such a catastrophe is to launch a deep-space mission with an on-board high-yield thermonuclear device against the incoming body. It would be most effective if it penetrated to the center of that body, either to disintegrate it into smaller and less dangerous fragments or to change its trajectory. If a special International Space Observation Service were organized for the detection of suspicious objects in deep space, then bodies such as the 1989 fc asteroid could be detected at distances which represented 5–6 months travel to the earth (according to the current state of the art in astronomy).

Spacecraft for this purpose could be designed and tested by the international team well in advance. The IFMC could be responsible for designing a very-high-yield thermonuclear explosive device which could be assembled and launched in a few weeks so that the incoming body could be intercepted far enough away. This International Asteroid Protection Service should therefore have ready "virtual" nuclear explosives within a zero-nuclear-weapon environment.

References

1. Theodore Taylor, "The Global Abolition of Nuclear Weapons" and Stanislav Rodionov "Verification of Compliance and Deterrents to Violations" in *Towards a Secure World in the 21st Century: Annals of Pugwash 1990*, eds. J. Rotblat and F. Blackaby (Taylor & Francis, 1991).
2. Steve Fetter and Frank von Hippel, "The Hazard from Plutonium Dispersal by Nuclear-Warhead Accidents," *Science & Global Security*, Vol. 2, No. 1 (Gordon & Breach, 1990).
3. Joseph Rotblat and Vitaliĭ Gol'danskiĭ, "The Elimination of Nuclear Arsenals: Is it Desirable? Is it Feasible?" in *Verification: Monitoring Compliance*, eds. F. Calogero, M. L. Goldberger, and S. Kapitza (Westview Press, 1991); Vitaliĭ Gol'danskiĭ, "UN Nuclear Forces? A Soviet Scientist's Proposals," *Izvestia*, No. 311, 9 November 1990.
4. Theodore Taylor, "Can Nuclear Weapons Be Developed Without Full Testing?" in *Verification of Arms Reductions*, eds. J. Altman and J. Rotblat (Springer-Verlag, 1989).
5. Dan L. Fenstermacher, "The Effect of Nuclear Test-Ban Regimes on Third-Generation-Weapon Innovation," *Science & Global Security*, Vol. 1, Nos. 3–4 (Gordon & Breach, 1990).
6. Ray E. Kidder, "Military-Significant Nuclear Explosive Yields," *FAS Public Interest Report*, Vol. 38, No. 7, September, 1985.
7. A. P. Vasiliev, N. K. Prikhodko, and V. A. Simonenko, "Underground Nuclear Explosions for Improvement of Ecological Situations" (in Russian) *Priroda*, No. 2, February, 1991.

Could Our Countries Become Allies?[a]

Q *You have said in the past that if we presume that Soviet-American dialogue is largely responsible for developments on this planet, then we can say that the relations between our two countries have priority value. Would you please elaborate on this?*

A: Many European and Asian countries have a strong say in international affairs today. On a number of issues their opinion is rated at the same level as the opinion of the United States or the Soviet Union. To shape a prudent international policy, both for the present and in the long-term perspective, it is worth proceeding from the premise that Soviet-American relations are of special value.

I want to stress that this is my personal view. I don't want anyone to misinterpret what I say to mean that the relations of the Soviet Union or of the United States with third countries, including their respective allies, are of a secondary nature. There are more things in the world that unite countries than that separate them. On the basis of my own experience and the economic and military potential of our two countries, as well as of the history of their confrontation, I consider the development of the relationship between the Soviet Union and the United States a top priority in foreign policy. By the way, upon the completion of his latest meeting with the President of the United States, President Gorbachev said at a press conference that "what we have been doing here will be useful, not only for the people of our two countries,

[a] Interview by Aleksander Ivan'ko, a correspondent for the newspaper *Izvestia*.

but for the world as a whole." So the Soviet Union and the United States can assume the role of pilots in the stormy sea of world politics. I think that everyone will benefit from this.

This very principle has been supported in the United States, although doubts were expressed concerning the possibility of adopting a corresponding joint declaration in the near future. Because so little time has passed since we regarded each other as enemies, it is rather difficult to regain an awareness of the closeness of our interests. But we should do this.

Q: *What led you to this idea?*

A: Like the ripples begun when a stone is thrown into water, Soviet perestroika has triggered cardinal change around the world and, in particular, in our neighboring Eastern European countries. The downfall of the "fraternal" regimes, which seemed indestructible, and the prospect that some countries may eventually withdraw from the Warsaw Treaty are creating an altogether new situation in Europe. At the same time, the idea of a common European home is gaining momentum. I envisage the merger of the two military blocs into one political bloc with the tentative name ESTO [European Security Treaty Organization]. Military confrontation will give way to a situation in which there is a single political bloc. In this respect a policy emphasizing the supremacy of the values of Soviet-American relations and recognizing the growing role of the United Nations is of prime importance. I see such a policy as the surest guarantee of security for our common European home.

The Soviet Union and the United States are the only countries capable of annihilating each other and all humankind several times over. We have finally realized this and have begun a process of disarmament. I hope that a treaty calling for a 50 percent reduction in strategic offensive armaments will be signed soon. And experts are working on a plan to reduce those arms by 90 to 95 percent, which would make the remaining nuclear armaments in the possession of the Soviet Union and the United States commensurate with those of the other nuclear powers. That may convince them to join the disarmament process. Military spending is a tremendous burden for our countries. For the United States it poses the danger of lagging behind the new giants in the economic race: a united Europe—with its center in a new Germany—and Asia and the Pacific region—whose center is in Japan, with a good chance of China joining the race. The emergence of such powerful forces can be a factor for either stabilization or for destabilization. If we proceed from the suggested policy, the chances of stabilization increase. In an era when the number of nuclear powers may secretly grow, a 100 percent reduction of nuclear arsenals is hardly prudent. But nuclear police forces can be set up under the aegis of the United Nations.

Before an organization like ESTO can be formed, we must deal with a very important issue—economic support for *perestroika*. An economically prosperous Soviet Union is a major condition for global stability. Conversely, an impoverished Soviet Union, beset by anarchy resulting from ethnic strife, would be a very dangerous, destabilizing factor. I think that worldwide support for Gorbachev is a matter of global importance. The approach of the United States to the complicated processes on the Soviet political scene should be based on the supremacy of the value of our relations. Pragmatism should prevail over emotions.

Q: *What is needed for your approach to Soviet-American relations to become a reality?*

A: There's no magic wand we can wave to switch from confrontation to cooperation. But the ways in which the transition can be made have already been outlined in discussions at the Washington summit. The main thing is strict verification. Not only mutual verification by the Soviet Union and the United States, but some sort of fire tower, a vantage point from which it would be possible to see whether a fire is about to break out somewhere. Both our countries have the maximum possible opportunities for this; we can become the core of a future international observation and detection system. Mutual verification builds mutual trust, which in turn facilitates verification. We have even started reappraising our terminology. The infamous "spy satellite" becomes the much-needed means of verification, a factor for stability. Within the framework of such a system of verification, it would also be possible to create a demilitarized zone across the whole of central Europe.

The proposed policy is also very useful for settling regional conflicts. Many conflicts arose from the policy that existed earlier: Everything is advantageous to the Soviet Union that is disadvantageous to the United States and vice versa. Sources of tension appeared periodically, and we and the Americans were only adding fuel to the fire by supporting the opposing sides. If we proceed from the premise that any conflict could damage our relations, then we will be much more active in extinguishing fires.

Q: *Your principle proceeds from the existence of a strong Soviet Union, but it is no secret that our country is being rent by the most diverse conflicts—from the economic to the interethnic. Some political leaders in the West, and indeed in the USSR, are even talking of the imminent disintegration of the Soviet Union. Don't you think that your doctrine may become less viable in such a situation?*

A: Even if worst comes to worst, Russia in itself—by virtue of its potential, its size, its resources, its people's talent and capacity for work, its contribution to the world's culture—will remain a great power if it stands firmly on its feet and is stable politically and economically. I think the question you are raising is this: Is this policy mutually attractive? It is obvious that today it is more attractive to us than to the Americans. But how advisable is it for the United States to pay so much attention to relations with a country that is on the verge of disintegration?

Another question also arises: Is there any need then to support perestroika? In my view, Gorbachev's perestroika and stabilization at home are equivalent to stabilization worldwide, whereas the road to anarchy will lead to the emergence of a seat of danger to the planet as a whole. It seems obvious to me that the United States and the West are vitally interested in the success of perestroika.

Q: *You have spoken of economic support for perestroika. What do you mean by that—regular financial injections into our economy?*

A: Several proposals are now being suggested. There is talk both of a new "Marshall Plan" for the USSR and Eastern Europe and of lend-lease deliveries in the mold of World War II. These are extreme measures, and it is far from obvious that they will be helpful in themselves. In the final analysis, we must straighten things out at home by ourselves. In my view, the United States should first do away with the discriminatory acts impeding the development of trade with the USSR.

Q: *Don't you think that your idea may encounter substantial resistance from the conservative forces in our country?*

A: I can't imagine any single policy that everybody in the country would approve of. At the same time, the policy in question is based on realistic tendencies in the development of the world today. The alternative is to continue to arm ourselves to the teeth and to revive the old conception of America as a potential adversary. This would mean quite a different policy, with the obligatory presence of the enemy image. That's the policy we lived with for the past 72 years, seeking external and internal enemies everywhere. If we deny the priority of values central to all humankind over class-based priorities, then we are bound to negate the policy I have proposed. So I imagine I'm going to have enemies.

Finally, I would like to emphasize once again that humankind is facing problems of a global scale, which not only both our countries, but also groups of countries and military-political blocs, will be unable to cope with. The world community must join hands and, utilizing the strength and prestige of the United Nations and other international organizations, it will, perhaps, find ways to solve the planet's problems. As for the role of the uniting and organizing principle, it could be assumed by the USSR and the United States—I envisage precisely such actions by them in the foreign policy arena.

UN Nuclear Force?
A Soviet Scientist's Proposals

Today I no longer believe in the feasibility of transition to a nuclear-free world during the remaining 10 years of the 20th century. Whether we like it or not, we will evidently be obliged to reconcile ourselves to the existence of nuclear weapons in the immediate future. But in what form and in what numbers? Does this mean that nuclear tests will also continue, to the detriment of the atmosphere—above all, the public and political atmosphere?

In the quest for an answer to these questions, I wish to advance some new proposals whose implementation could, in my view, make the world safer and bring the nuclear-free era closer. But first a few words about facts that are of concern to everyone—above all, the recent underground nuclear explosion carried out on Novaya Zemlya by the Soviet Union.

In my view, it rightly elicited a sharply adverse response from public opinion both in our country and abroad. True, mention must be made of our 19-month unilateral moratorium during 1985–1987 and the more than one-year silence at Soviet nuclear test sites from October 1989 through October 1990. Yet during the same time seven explosions have been carried out on the U.S. test site in Nevada (during 1987–1989, according to data cited in *Izvestiya* No. 274, there were 39 explosions in the United States and 47 in our country). Three other nuclear countries—Britain, France, and China—also have "a finger in the pie," as the saying goes.

Yet, in connection with the October 24 test on Novaya Zemlya, I recall the words which Joseph Fouche spoke after Napoleon ordered that the Duke d'Enghien be captured outside France and shot: "It was worse than a crime—it was a blunder." It is hard to describe as anything else an action which further exacerbates relations between the leadership of the USSR and the RSFSR [Russian Soviet Federated Socialist Republic] and creates a factor of inimicability toward M. S. Gorbachev in Scandinavian countries on the eve of his triumphal Nobel visit to Oslo and Stockholm.

However, I would like to look not to the past but to the future. I will try briefly to describe the state of affairs with regard to nuclear weapons, as I see it in the light of the present-day picture of the world.

I am sure that no real danger now exists of a "hot war" between one or another (or several) of the five current nuclear powers—in any combination.

I am also convinced that not one of these five powers, which have seen for themselves the awful destructiveness of nuclear explosions, will be the first to use nuclear weapons either against each other or against any third country, regardless of whether or not a given nuclear country has proclaimed corresponding declarations (although, of course, such declarations are highly desirable!). But under such conditions, from my viewpoint, there is no point in nuclear powers' having in their arsenals multiple-charge systems for the delivery of nuclear warheads (so-called MIRV's), nuclear charges with a megaton yield, or any tactical nuclear weapons at all—the majority of types of first-and second-generation nuclear weapons currently in the arsenal. Work on creating third-generation nuclear weapons is all the more pointless.

At the same time it has to be acknowledged that some other countries in the world already are—or could become in the very near future—"underground" possessors of nuclear weapons, for example India, Israel, the Republic of South Africa, Pakistan, Iraq, North Korea, Argentina, Brazil, and others—and they will not need any preliminary tests to be fully sure of the effectiveness of such weapons. Nor do I rule out the possibility of nuclear weapons falling into the hands of organized criminal groups—let me remind you in this context that the smallest models of modern nuclear weapons weigh only approximately 50 kg, have a diameter of less than 20 cm, and explode with a yield equivalent to approximately 1,000 tons of TNT. What is to be done about these dangers?

Missiles with a very high degree of accuracy and great depth of penetration, not even necessarily carrying nuclear charges (and, if they do, ones of minimal yield), appear the most effective to give a warning or to punish specific architects of aggression in the "nuclear underground."

The U.S. and our own armorers now usually substantiate the need to continue nuclear tests with reference to the need to enhance the reliability and safety of nuclear weapons. What they understand here by reliability is the condition that the weapon should operate efficiently whenever necessary, and what they mean by safety is that it should not operate without the involvement of people initiated into the secrets of safety and code systems, "by itself," or in criminal hands, even under any extraordinary conditions.

Thus, it is a question of seeking an absolutely reliable guarantee against any unpremeditated, chance catastrophic events. It can be considered that in all models of existing nuclear weapons both in the United States and in the USSR the possibility of an unpremeditated nuclear explosion is virtually ruled out—even if the chemical explosive surrounding the nuclear core were to explode. However, another danger exists—that such an explosion would result in the discharge of very tiny particles of highly toxic, radioactive plutonium into the atmosphere. And the quantity of plutonium in an ordinary nuclear charge is only three or four times less than the total quantity of this substance ejected during the Chernobyl disaster (approximately 20 kg).

I will repeat here what I wrote repeatedly during our 1985–1987 moratorium in opposing the U.S. nuclear tests (see, for example, *Izvestiya* No 335 for 30 November 1986): Full-scale tests involving nuclear explosions are certainly not essential to check the reliability and safety of nuclear weapons, for it is possible to manage with a sum total of checks of individual components, as well as "mathematical experiments" on computers and so on, but without exploding a nuclear charge.

So, a nuclear-free world remains a goal for the future. For the time being I see the creation of an international nuclear police force under UN auspices as an extremely necessary and inevitable stage on the way to it, when the Security Council and its Military Staff Committee must collectively have at their disposal not only "paper" but also real "buttons" for the immediate and effective use of this force against any nuclear aggressor. Such a UN nuclear force must be created against the background of a simultaneous and very strong reduction in the nuclear arms of the United States and the USSR (to approximately 1,000–2,000 charges), the guaranteed and controlled full destruction of chemical and biological weapons, a sharp reduction in conventional arms, progress on the way to creating an all-European and an Asia-and-Pacific security system, political and economic cooperation, the elimination of existing interregional conflicts, and the prevention of new ones.

The UN nuclear force would have to be preserved for a while even after the total elimination of all national nuclear arms, and only then be eliminated under strict international control.

As for underground nuclear tests, if there is no way the United States can renounce them, let us reach agreement on conducting all such tests jointly, gradually reducing their frequency and the upper limit for the permitted yield of an explosion. Where? In Nevada, for example, since the authorities in that state gave their consent to the functioning of a nuclear test site there. I believe that in this way we could really help to strengthen mutual trust, bring substantially nearer the complete cessation of nuclear tests, strengthen the Treaty on the Nonproliferation of Nuclear Weapons, and make potential nuclear aggressors listen to reason, in other words create the necessary preconditions for transition to a nuclear-free world.

The World Wants Clarity and Reliability (When It Concerns Nuclear Weapons from the Former USSR)

There is perhaps no other problem which worry the people of all the continents as much as the fate of nuclear weapons in the former USSR.

The Bulletin of the Atomic Scientists, published annually in the U.S., has a symbolic clock standing at nearly midnight on its cover . After the conclusion of the Soviet American Treaty on Intermediate and Short Range Missiles the clock was put back from three minutes to seven to midnight, and since April 1990 it stood at ten to midnight. It seemed, after the signing of the Soviet American Treaty on Strategic Arms Limitation on July 31, 1991, that one could boldly continue the counterclockwise motion of the minute hand and, indeed since December 1991, the minute hand was moved back again—to seventeen minutes before midnight, stronger than ever before. But meanwhile the world has encountered new, unforeseen threats coming exclusively from the side of the former USSR, armed not only to the teeth but literally beyond measure.

Only Russia Has a Right to Remain a Nuclear-Weapon Country in CIS

At first during the three days of the August 1991 putsch it was not clear who owns the "nuclear button", and whether it had fallen into the hands of the conspirators. There were indeed some frightening signs. It is enough to reread some articles from pre-putsch newspapers written by high-ranking scientists and officials bound up with the military industrial complex, let alone the first diplomatic documents of the August junta.

Now the world is still uneasy though for different reasons. There is no yet adequate clarity in distribution of functions of the former USSR among the members of the Commonwealth of Independent States (CIS) on which territories nuclear weapons are deployed.

Russia has been recognized to be a legitimate successor of the USSR as a permanent member of the UNO Security Council with the right of veto. It is necessary to fix, as soon as possible, the analogous status of Russia in the two most important international treaties which serve as reliable safeguards against the threat of nuclear war. These are the Nuclear Weapons Non Proliferation Treaty (NPT) (1968) and the Treaty on Limited Nuclear Test Ban (LTBT) in three Media (in atmosphere, outer space and under water). In both these treaties the USSR, along with the USA and Great Britain, is one of the three depositary countries (keeper of ratified certificates) which possess the right to veto any amendments.

The number of countries participating in the NPT was 141 in 1991. Last year France and China joined it. This treaty clearly formulates the rights and obligations of two categories of participants—nuclear and non-nuclear countries. It defines nuclear countries as the ones which produced and exploded nuclear weapons before January 1, 1967. Initially there were five such countries: U.S. (1945), USSR (1949), Great Britain (1952), France (1960) and China (1964). At present of all the CIS members only Russia has a well-grounded right to claim to the role of a nuclear country. Though during the first five years (1949–1954) all Soviet tests were held exclusively in Kazakhstan, production of nuclear weapons was always the monopoly of Russia.

It seems urgent now that the transfer of all rights and obligations under the NPT to Russia should be officially formalized and all the other CIS members become party to the treaty with the status of non-nuclear countries. It is worth noting here that according to Article 5 of the relevant Alma-Ata document of December 21, 1991, the Ukraina and Belarus' have already assumed responsibility of becoming parties to the treaty in such capacities, but Kazakhstan has not.[1]

[1]On May 19, 1992, during his visit to the U.S., Kazakhstan's President Nursultan Nazarbayev declared the willingness and determination of his Republic to join officially the NPT under the status of non-nuclear country. This news marks an extremely important positive step in the maintenance and strengthening of non-proliferation regime.

The U.S. and Great Britain May Reconsider Their Position on Nuclear Tests

Significance of official establishment by a treaty of the future status of former Soviet nuclear weapons is so great that it seems quite timely to think of convening an urgent extraordinary conference of the countries which are parties to the NPT (usually convened every five years, in 1995 the further status of the treaty should be determined). Besides this, such a conference will provide a good opportunity for settling or raising again, this time more persistently, the question about the comprehensive test-ban treaty (CTB).

This comprehensive ban could be legitimized as an amendment to the LTBT of 1963. To the already mentioned three media banned for tests, the fourth, underground, can be added. The present number of the parties to the treaty of 1963 is 118, of which 104 also signed the NPT. So to convene the conference for consideration of the amendments to the LTBT of 1963, the initiative (or consent) of 60 countries could be enough, and is easily achieved. However the passing of the amendment on CTB can be completely blocked by the veto of the U.S. or/and Great Britain. Still in the present international situation these countries may reconsider their adamant positions on CTB. It seems that the political gain of their consent to such a ban could far exceed rather disputable technical arguments in favor of continuing tests [1]. In case the U.S. will persist on their continuation, I'm repeating my earlier proposal [2]: let us conduct not only American and English but also Russian nuclear tests exclusively in Nevada; moreover, decrease their frequency (to, let us say, no more than two annually) and strongly limit their upper yield (e.g. by few kilotons).

By the way, growing scientific and technological cooperation of American (generally Western) and Russian nuclear scientists, their joint efforts in solving the most promising problems of fundamental science, along with the experience gained in these countries and sophisticated equipment of their leading atomic centers, may be a reliable shield against migration of CIS qualified scientists to countries like Iraq or Libia.

The change of the position of the U.S and UK towards the CTB seems to be even more probable now, after France has announced its decision to join the unilateral Russian moratorium for underground nuclear tests—now valid until October 1992. Continuation of tests by the U.S. and UK places Russian President Boris Yeltsin in quite an ambiguous position. Before the dissolution of the USSR Yeltsin announced the complete cessation of underground nuclear explosions at Novaya Zemlya (as Kazakhstan's President Nursultan Nazarbayev decided for Semipalatinsk) and complete elimination of this test site. However at the end of February 1992, under the strong pressure of the

Russian military industrial complex, Yeltsin ordered the preparation of the Novaya Zemlya test site for additional tests to resume—if the moratorium is not reinforced by the U.S. and UK.

This decision of Yeltsin's was met by the strong protests of northern European countries and of representatives of small-numbered folks of Northern Russia at the April session of the Congress of People's Deputies.

It would be worthwhile to emphasize the extreme significance of the support of B. Yeltsyn by immediate joining of the US and UK the Russian and French underground test moratorium [3] (see Supplemento 1).

The Way to Eliminating Tactical Nuclear Weapons

To conclude I'd like to add a few words about the most serious threat contained in tactical nuclear weapons which can be used without authorization of the leaders of CIS countries and armed forces, and thus can easily fall into irresponsible or even criminal hands. Even if a malefactor cannot conduct a nuclear explosion, chemical explosion of the shell surrounding each charge will blow out into the atmosphere highly toxic radioactive plutonium in the amount of 3–4 times less than that in Chernobyl.

We discussed the urgency of total elimination of tactical nuclear weapons with our American colleagues in September, 1991 at the 41st Pugwash Conference in Beijing. Declaration of the Pugwash Council anticipated the September initiatives of Presidents Bush and Gorbachev and was supplemented by the attached document [4] (see Supplemento II).

Let us hope that within the few following years (maybe even months) the main suggestions of this document will be realized as well as political measures described at the first parts of this presentation.

References

1. Sidney D. Drell "Merits and Risks of More Underground Tests" *Physics Today*, **44**, 9–12, November, 1991.
2. Vitaliĭ Gol'danskiĭ "UN Nuclear Force? A Soviet Scientist's Proposals." *Izvestiya*, November 9, 1990.
3. Supplemento I. George Perkovich. "To Bolster Yeltsin, Stop U.S. Testing" *Los Angeles Times*. News Fax. Moscow Edition. April 24, 1992.
4. Supplemento II. Memorandum attached to the letter of F. Calogero and J. Rotblat to Presidents G. Bush and M. Gorbachev dated September 23, 1991.

Statement of Russian Scientists

It is well known that in recent years there have been widespread and embittered attacks on scientists and science in general, in which science is often accused of being responsible for all kinds of disasters. We are compelled to acknowledge that such obfuscation is particularly widespread in our country, where attacks on science have been used not only to discredit our intelligentsia but also to create political tensions internally and internationally, to produce an atmosphere of xenophobia, and to generate hostility and mistrust within the world scientific community.

These were the aims, in our opinion, of the book "Special Tasks"—written by the long-time spy Pavel Sudoplatov (in co-authorship with his son Anatoli and two Americans, Jerrold and Leona Schechter) and recently published in the United States. This book has received widespread publicity because of its sensational allegations of espionage activities on behalf of the Soviet Union by some of the most distinguished American Manhattan Project scientists; it also alleges, on a related note, that the most prominent scientific leaders of the Soviet atomic- and hydrogen-bomb projects were too incompetent to have succeeded in these projects without the help of information stolen by spies.

These latter allegations were put forward recently in our country in the newspaper "Sovershenno Sekretno" ("Top Secret," June 1994), in a television show with the same title (June 4, 1994), and in the newspaper "Moskovskii Komsomolyets" (June 29, 1994). In this open letter of response, we—Russian scientists—would like to make two statements on behalf of our friends and colleagues.

First, we reject categorically the absolutely unsubstantiated, malicious, and provocative allegations that Robert Oppenheimer, Niels Bohr, Enrico Fermi, Leo Szilard, and George Gamov were sympathizers, accomplices, or assistants in the spying activity of Klaus Fuchs. Sudoplatov's book does not contain a single fact to support the validity of these accusations. We fully endorse the view expressed in the editorial, "Unsubstantiated Charges of Treason," which appeared in the April 28, 1994 issue of *Nature*.

Second, we think that the history of the creation of the atomic bomb first tested in the USSR on August 29, 1949, is correctly described in the articles of Yuli Khariton and Yuri Smirnov in "Izvestiya" (December 8, 1992) and *The Bulletin of the Atomic Scientists* (May 1993). These articles accurately note the information value of the materials passed by Klaus Fuchs to the Soviet intelligence network and delivered to the leader of the Soviet atomic project, Igor Kurchatov.

Khariton and Smirnov also describe the reasons that led to the decision to explode, in the first Soviet test, a copy of the American implosion-type bomb that had already been tested twice before, at Alamogordo and in Nagasaki. We should add that under the conditions that existed during the totalitarian regime of Stalin, a failure of the first test of a Soviet atomic bomb would certainly have led to harsh punishment for all of Kurchatov's team and to the total devastation of Soviet physics (as happened shortly before, in 1948, to Soviet biology). Such a devastation would have begun with the condemnation and prohibition of relativity theory and quantum mechanics, ostensibly because they were the poisoned tools of ideological saboteurs. The atomic test in August 1949 in fact saved Soviet physics.

As is well known, it was only two years later—in the fall of 1951—that the Soviet Union successfully tested two atomic bombs of original domestic design. In the early development of thermonuclear weapons, moreover, Soviet scientists out-distanced their American colleagues.

We resolutely reject, therefore, the malicious and provocative attempts of reactionary, antidemocratic, antireformist, and anti-intellectual forces in Russia to disparage the scientific and technical achievements of the Soviet scientists who worked on the creation of atomic and thermonuclear weapons, and to accuse them of being mere imitators who blindly followed the prescriptions of information pilfered by intelligence services.

SIGNED

Academician Zhores I. Alferov
Academician Aleksander F. Andreev
Professor Arkadii A. Brish
Academician Aleksander Ye. Chudakov
Academician Aleksander V. Fokin
Academician Vitaliĭ I. Gol'danskiĭ
Academician Yuri A. Ossipyan
Academician Aleksander N. Skrinskiĭ
Academician Mikhail A. Styrikovich
Academician Valerii I. Subbotin

transmitted by Academician Vitaliĭ I. Gol'danskiĭ
edited by Dr. Victor Rabinowitch and Prof. John P. Holdren

POLITICS

While being a Member of the USSR Congress of People's Deputies (from May 1989 till the decomposition of the USSR at the end of 1991) I was elected to the Committee for International Relations of USSR Supreme Soviet.

My lively interests in the problems of our foreign policy, support of the Gorbachev-Shevardnadze-Yakovlev line (directed to the complete elimination of the remnants of cold war), and my aspiration to propagandize tighter USSR-U.S. rapprochement are reflected in the selections of this section.

The triad of articles in *Izvestia* (June 1990–January 1992) should be emphasized. A manuscript of the first article was delivered by me to President Gorbachev in connection with his visit to the U.S. in May 1990.

Unfortunately, space limitations deprive me of the possibility to include my political speeches at the four sessions of the USSR Congress of People's Deputies. Inter alia, I strongly supported the immediate election of Mikhail Gorbachev as the President of the USSR (March 1990), and later expressed my deep concern and anxiety in connection with the resignation of Foreign Minister Shevardnadze and his warnings about the forthcoming offensive of reactionary forces (December 1990). All these speeches were published in *Izvestiya*.

An Academician Demands Publication of the Military Budget[a]

Editor's Note, *Le Matin:* Disarmament provokes debates in the West. In the USSR the debate is in full swing and pits the opponents and partisans of Mikhail Gorbachev against each other. Academician Vitaliĭ Gol'danskiĭ is one of the scientific partisans of rapid disarmament.

This physicist, who heads an important institute of nuclear physics and chemistry, demands, in particular, that the Soviet military budget be made public.

Vadrot: *Are you optimistic about the results of the Reagan-Gorbachev meeting?*

Gol'danskiĭ: Yes, I am both very interested and fairly optimistic. The meeting has been prepared for very well and should lead to good results.

Vadrot: *What do you think about the reservations which Europeans seem to have regarding this denuclearization of European territory?*

Gol'danskiĭ: I understand their worries, and this is a subject which I often bring up with my French colleagues and with those of the other countries concerned. It is reasonable for them to be somewhat anxious. This shows that we cannot stop here, that we must tackle other problems of disarmament, beginning with conventional arms.

Vadrot: *But won't the Soviets have the possibility, finally, after the disappearance of the European missiles, of reaiming other missiles not involved in this first agreement?*

---

[a] Interview by Claude-Marie Vadrot.

Gol'danskiĭ: You are correct, this is a real problem. This must also be a matter for monitoring and verification. It must be impossible for someone to cheat.

Vadrot: *Will dismantling the missiles be expensive?*

Gol'danskiĭ: Yes, because it is a complicated operation. It is necessary to first disentangle the propulsive part and the combustible part. Then the problem of the nuclear warhead must be dealt with. I know that there are French scientists with the technical skill to manage this, to transform the combustible nuclear material into material for civilian use.

But ultimately the problem is as much political as technical, and it requires setting up an international collaboration, a working group to solve this problem. A peaceful competition can be started.

Vadrot: *Will disarmament ultimately lead to a decrease of the Soviet military budget?*

Gol'danskiĭ: Yes, that is coming. The productive forces of a country are necessarily limited, and it is in our interest to cut military spending and raise civilian spending. This is valuable not only for us... All humanity will profit from it.

Vadrot: *It would be good if your military budget were public...*

Gol'danskiĭ: You are quite correct; I wish it would be released, and calculated in the same terms as those of the western countries! It is at the same time an economic problem and a political problem. Such open publication and calculation are necessary to have truth in pricing, to understand what are the real costs; and democratization implies truth on this subject. Openness must also apply to military affairs.

Vadrot: *How do you see disarmament coming about?*

Gol'danskiĭ: It is necessary to be very strict regarding the question of mutual verification. This verification must be accurate, constant, while little by little we come to trust each other. Mistrust must become a thing of the past: it is necessary to verify and verify again, including in the area of chemical weapons. The countries involved must open themselves to others, and scientists can be the guarantors of this initial stage in developing mutual trust.

Moreover, the International Pugwash movement which I participate in is preparing a blue book on this topic, and we are going to discuss it, among scientists, in Geneva in January, in connection with chemical weapons.

Vadrot: *Is there any hope of a reduction in conventional weapons?*

Gol'danskiĭ: Yes, it is enough to find some political solutions and, I repeat, to set up verification procedures. But this should happen quickly, too.

Vadrot: *Will this summit not lead to any surprises, or do you think that Mikhail Gorbachev has other propositions or other initiatives in reserve?*

Gol'danskiĭ: It is quite possible that other things will be announced. Everything is possible once things are set into motion, even though it is still difficult, because every suggestion must have a symmetry. But I indeed think that the time to make other plans is now or never.

The World Has Become Extremely Small[a]

Editor's Note: Academician V. I. Gol'danskiĭ is well known as an active propagandist for a multilateral relation between scientific-technological progress and the new political thinking. Our correspondent Stanislav Pestov discusses this with him.

Pestov: *Vitaliĭ Iosifovich, I heard you say something that at first glance appeared to be a paradoxical evaluation of the role of spy satellites. Could you speak about this in more detail?*

Gol'danskiĭ: Our minds are literally cluttered with stereotypes of the past. People of every country in the world are consumed with the desire to accumulate more and more weapons of greater and greater destructive power at any price. Given this situation, spy satellites represent a threat precisely where secrecy is viewed as the "front line of defense." It must be said that deterrence based on the threat of retaliation has in the past played a positive role, mainly because it broke the American monopoly on nuclear weaponry. Europe has gone for more than 40 years without a war.

But the swollen armed forces is a very expensive psychological weapon; it can command fear, but not safety.

Pestov: *It would seem that the concept of mutual deterrence by means of the power of arms must give way...*

Gol'danskiĭ: ...to the concept of mutual trust. But I would characterize any attempt to switch directly from deterrence by arms to deterrence by trust as utopian. An intermediate stage is needed, and we have essentially already entered it. This is the stage of "verificational deterrence," based on verification, i.e., on extremely strict mutual monitoring and control. It is in full accordance with the principle of glasnost in international politics. The world

[a] Interview by Stanislav Pestov.

has become extremely small and open, in principle, everything is subject to monitoring and verification owing to the development of outer-space exploration, microelectronics, and informatics.

From this point of view I consider the "spy satellite" an outdated stereotype, something akin to the "image of the enemy" stereotype. Actually, at the present time, satellites, like everything else which aids observation and verification of adherence to treaty obligations, stabilize the world. But the allocation of enormous means to the Stealth Bomber invisible to radar is a destabilizing factor.

Pestov: *Reduction of the armed forces and weaponry, which our country has recently begun unilaterally, also stabilizes peace.*

Gol'danskiĭ: And it in no way has weakened us. When we demonstrate openness, when we confirm it by specific actions, we also generate trust. It creates positive feedback. Verification generates trust, and trust broadens the possibility of verification. Conversely, the secrecy typical particularly to both our bureaucracy and the bureaucracy overseas generates suspicion... . For example, who believes that each year we are spending a sum of order only 20 billion rubles on military purposes?

Pestov: *The American military expenditures exceed 300 billion dollars a year, and to preserve military parity we must be spending much more than 20 billion, even allowing for the fact that our army is not hired, i.e. not paid.*

Gol'danskiĭ: According to the data of the American intelligence sources, which is the only publicly available data on this question, we are spending more than 220 billion dollars for military purposes.

This item in our list of expenditures is quite strange. The other items can be estimated more accurately, in spite of the absence of any order in the prices, but here... . However recently it was announced that our military budget would be cut by 14.2%. This number has three significant figures, but does that mean it can really be used for calculating? These bureaucratic games are not only comical, they are also dangerous at the present time, when establishing trust should be becoming the normal behavior. The removal of senseless restrictions on the information supplied to the Soviet people, the complete declassifying of all information which does not contain state or military secrets, should become one of the main priorities of the People's Deputies and the commissions of the Supreme Soviet. This requires serious and precise legislation, which must be based on the presumption of nonsecrecy: anyone wishing to classify some information must prove that it is really necessary to do so.

I propose that the deputies should do everything necessary to transform the economics from one pursuing armament to one pursuing disarmament, while preserving our military strength only at the level dictated by defense doctrine.

Part of the huge debt of the developing countries includes many billions of dollars owed us for weapons we have shipped to them. But even Lenin himself had emphasized that the principal assistance which we can give to national liberation movements is not arms shipments, but the inspiring example of the successful creation of a new society. Burden of military expenditures are harmful to both developing countries and developed ones.

At present, the total strength of the nuclear explosives stockpiled around the world amounts to 10–15 thousand megatons. That is three tons per person!

I just had the thought: if all this annihilation energy were transformed (of course, this is physically impossible) into energy to support life, say, into calories of butter, 300 kg of butter would be obtained for each inhabitant of the planet. Or something like three tons of meat. If its annual consumption is 60 kg, there would be enough meat for each inhabitant for 50 years!

Pestov: *Vitaliĭ Iosifovich, would a "conversion" transforming nuclear explosives into fuel for nuclear energy be possible?*

Gol'danskiĭ: Yes, it is not too complicated technically. For instance, weapon-grade uranium can be used for the enrichment of natural uranium, which would be considerably simpler and cheaper than the present enrichment methods. If all these stored 15,000 megatons were supplied to nuclear power plants, the total power of which in the entire world is 400 million kilowatts, these plants could operate on this "converted" fuel for two years. And if fast-neutron fission is also used, for example, as proposed in France, to create reactors operating on mixtures of molten salts of plutonium and thorium, the energy available from nuclear explosives would be sufficient to provide nuclear energy for 20 years.

Pestov: *The fact that such huge amounts of nuclear explosives have accumulated is apparently the fault of not only "the other" side, but also of our scientists.*

Gol'danskiĭ: If at the end of the 1950s we had had the level of democracy and glasnost existing today, I think that the sad event of 1961 would have never occurred: our country was the first to violate the three-year-long unproclaimed moratorium on all forms of nuclear weapons testing, which both we and the Americans had adhered to. At that time there was essentially only a single person who protested: Academician Sakharov.

A bit later, also mainly through our fault, no agreement could be reached on a complete ban of nuclear testing: we rejected strict mutual monitoring, considering it the same as espionage on our territory. An excess of suspicion and enforced secrecy are what lead to such pernicious errors.

Pestov: *Vitaliĭ Iosifovich, in 1988, you were the only representative of science belonging to the USSR governmental delegation at the Third Special*

Session of the United Nations on Problems of Disarmament, and you have also participated in many international scientific meetings. Did you encounter situations at such meetings where even far-from-new materials from the foreign scientific literature were unknown to our participants?

Gol'danskiĭ: This has no longer happened in the last year and a half. Earlier it did happen quite often: we did not know about articles from the most widely read scientific journals of the United States and Britain, even when these articles contained arguments supporting our views on various current problems. And the reason for this was always simple: someone wanted to appear extremely vigilant, and the matter was carried to an absurd level. Our libraries were receiving foreign journals from which some unknown bureaucrats were cutting out these articles and even crossing out their titles in the table of contents... .

Pestov: *You are chairman of the Soviet Pugwash Committee, Vice-Chairman of the Committee of Soviet Scientists for Peace, Against the Nuclear Threat, Vice-Chairman of the Board of Directors of the All-Union Society Znanie (Knowledge), and much, much more. Does Academician Gol'danskiĭ have any private life, a favorite pastime, time for sports and family life?*

Gol'danskiĭ: I try to find time for walking and swimming, and I love and collect records and CDs, the latter mainly classical music.

Pestov: *Our readers are interested to know what an Academician's salary is.*

Gol'danskiĭ: For the rank of Academician I am paid 500 rubles. As institute director I get paid another 600 rubles.

Pestov: *But you are still the chief editor of the Soviet journal "Chemical Physics," a member of many national academies, and on the editorial staff of many foreign and international journals...*

Gol'danskiĭ: For all journals the work (and it is not negligible) is wageless. And memberships in foreign academies and societies are simply honorary titles.

Pestov: *It is clear that your efforts are applied in a fairly broad sphere. This is probably why you were proposed as a candidate for People's Deputies of the USSR by so many public organizations. Could you list the main points of your program?*

Gol'danskiĭ: To some extent they have been touched upon in our conversation. You asked and I answered the questions which people are most interested in. They form the basis of my platform. This must probably be true of every candidate for deputy.

Politics, Science, and Glasnost

The word "glasnost" no longer needs to be translated. Perestroika has put this concept into international orbit. Thus, the word "Sputnik" has acquired a companion. But instead of playing with words, we should compare the meanings of these two notions, one of which is a symbol of scientific-technological progress, while the other describes political renewal and democratization. Does glasnost have any serious, objective basis, independent of the goodwill of the "authorities"? Or can everything end unexpectedly? And if this does happen, what will it mean for us?

Of course, time passes very quickly: it seems it was only yesterday that we were arguing about whether or not glasnost was necessary, what should be its limits, and today we—along with many tens of millions of television viewers, radio listeners, and newspaper readers—have become the witnesses of an event as stunning in its openness as the Congress of People's Deputies of the USSR. However, this question still troubles many... .

All the arguments, sometimes extremely subtle and lengthy, about what is glasnost, how much of it should there be in our society, whom should it concern, and whom (or what) should it not concern, they all seem to me only to show from what a long, dark tunnel we should emerge into the light. Of course, all these discussions are, so to speak, historically justified, but I still think it amounts to arguing about nothing. We are participating in an objective, historically irreversible process. Its guarantee is the development of science itself.

It is not without reason that it is said that technological existence outdistances human consciousness. This is primarily the exponential growth in the flux of information, its variety, volume, and speed of retrieval, the improvement of the quality of the recording, storage, and transfer of information, for example, of such fundamental parameters of modern computers and other information systems as operative memory and speed.

"The world is moving," says Gianni Rodari. It is moving, whether we like it or not, and, in addition, our world has changed dramatically! It has become different, and it is important to understand this! And if we think about our still—alas!—rare photocopying machines, which continue to be locked up behind steel doors, we must immediately recognize what a threat is posed to science and the very intellect of our citizens. Several years ago Academician Roald Z. Sagdeev said to a general assembly of the USSR Academy of Sciences: "As long as we continue to view a Xerox machine as a class enemy, we shall continue to lag behind in science and technology." Unfortunately, the view that a Xerox machine is an unclean force has been manifested even at the Congress of People's Deputies of the USSR.

The special role of the photocopying machine lies in the fact that the scientific literature, in spite of its very strong "branching" and specialization, is becoming more and more accessible to scientific workers. The Xerox is becoming one of the most important tools determining the effectiveness of scientific information and the labor productivity of scientific workers, and it is becoming the most important technical tool in libraries (which here are poorly equipped!). Complete and rapid information transfer is essential for the most rapid and efficient development of science.

(I am not speaking here of the extreme importance of a knowledge of foreign languages, primarily English. The situation here is improving, though slowly, and we shall "optimistically" assume that by the end of the twentieth century many millions of Soviet people, including scientific workers, will be able to assimilate information in foreign languages—not only in written, but also in spoken form.)

Now new forms of international communications are appearing: for example, there are various data banks with automated inputs and outputs for subscribers. International radio broadcasting will in the future be eclipsed by satellite-based international television broadcasting. The personal computers which are becoming more and more widespread (so far mainly in the West) are not only devices for storing and transforming information, but they also propagate it.

In a word, the world is now becoming open like never before. It is pointless to argue whether this is good or bad. It should be understood that openness is becoming the means of existence of our civilization, the guarantee of its survival and progress. Therefore, democratization and glasnost objectively go hand in hand with scientific-technological progress, supporting each other.

And this means that it is impossible to solve one problem separately from another: e.g., to first achieve successes in science and technology, and then begin to develop glasnost, say, begin making photocopying machines accessible to all. Democratization and scientific-technological progress are organically related notions. Accordingly, it must be concluded that even if we are faced with only—I stress the "only"—the problem of developing the productive forces of the country and improving the economics, this will require a well balanced, clear, logical information policy which is rationally regulated. At present this is a very valuable commodity, the gold reserve of the country, the key to its power and success. Therefore, each of our citizens must be as informed as possible, he or she must be made a very active recipient of international and domestic information flow. This is the natural development of the postulates of the new thinking, the guarantee of the active participation of people in politics and economics, the route to economic regeneration.

If someone suddenly struck at our ability to keep informed, at glasnost, this would amount to the suicide of our society.

I am convinced that the concept of the informational society being created in the world must also apply to the politics of national security. To begin with, it should be noted that the concept of nuclear deterrence has become outdated, both technologically and morally. Of course, it must be admitted that at one time it acted as a brake preventing the outbreak of war between East and West, but today it has become a threat to the existence of humankind. The danger of a conflict arising accidentally has arisen, and the fate of humanity is controlled by computers, which can give the false signal.

We are alive largely owing to deterrence, as they say in the West. But how valuable is such a life, the life of nuclear hostages? Is it really impossible to think of anything better?

At present there is intensive and rapid development of technology which makes it possible to watch everything happening on the surface of our planet (and sometimes underwater or underground) from outer space using reconnaissance satellites equipped with all sorts of sensors. The number of stars on a military shoulder strap can now be counted from ethereal altitudes. This is a fact which inspires horror in the priests of secrecy. But can't we see in this something good, the basis for a qualitatively new and secure form of deterrence? For a reliable system of mutual verification? Isn't this system already being used to monitor adherence to the ICBM treaty? I am sure that nuclear deterrence must be replaced by verificational deterrence based on mutual control and monitoring.

Such methods are very effective, as is easily seen if one considers the opposite situation. After all, this is shown by our recent history: we were deprived of a great deal because we underestimated and feared these methods. From 1958–1961 the United States and the USSR adhered to an implicit moratorium on all forms of nuclear weapons testing. Unfortunately and to our shame, it was violated by our side. The excuse given was that France was continuing to test its weapons. Academician Sakharov spoke out strongly against this decision. At that time we were very close to a complete ban on all testing, but the chance was unfortunately lost. The treaty between the USSR, the United States, and Great Britain concluded in 1963 banned testing only in three media (earth, water, and the atmosphere). It was not possible to significantly limit underground tests because, among other reasons, we at that time were completely against any monitoring, considering it a form of espionage, a violation of the principle of sovereignty.

Incidentally, with regard to nuclear testing: I think that the unilateral moratorium to which we adhered at one time did not weaken us at all. On the contrary, it improved our image abroad and strengthened, if you will, our negotiating positions, and it should not have been stopped. The principle of parity has become outdated, ruinous, and senseless. Parity in nuclear weapons is doubly senseless. The inertia must be overcome.

One often hears the question: is the idea of verificational deterrence shared by our foreign colleagues? Let me give one example as an answer. At the beginning of last year I was able to spend some time in the United States, where I had very interesting meetings at a number of political science centers, for example, at Harvard University, or, more precisely, one of its research centers, which is headed by the well known biochemist and politologist Professor Paul Doty, then at the Herbert York's Institute in San Diego, and at the Los Angeles political science center headed by Michael Intriligator. I went everywhere with reports which were in general devoted to a single topic: scientific-technological progress and the new political thinking. Of course, the idea of verificational deterrence was somehow present in all these presentations, and it won new adherents.

I was many times forced to listen to how all our misfortunes—the ecological threat, the existence of weapons of mass destruction, of new illnesses and viruses—were caused by science, scientists, and their blind curiosity... I certainly cannot agree with this. As is well known, the creative seeking of scientists goes hand-in-hand with ideology and politics, with their prejudices and biases.

Let us recall that Bertrand Russell and Albert Einstein turned to humankind with their manifesto more than 30 years ago. Were these scientists listened to? After all, if one thinks about it, their ideas are the alphabet of the new thinking with all its features. Let us recall: "We have to learn to think in a new way. We have to learn to ask ourselves, not what steps can be taken to give military victory to whatever group we prefer; for there no longer are such steps; the question we have to ask ourselves is: what steps can be taken to prevent a military contest of which the issue must be disastrous to all parties."

"There lies before us, if we choose, continual progress in happiness, knowledge, and wisdom. Shall we, instead, choose death, because we cannot forget our quarrels? We appeal, as human beings, to human beings: Remember your humanity, and forget the rest"

When he became acquainted with the text, Nikita S. Khrushchev took it as an affront to himself and to socialism. Our leader, whose era I on the whole remember warmly, was certain that only capitalism would perish in the probable upcoming war—after all, it was "historically doomed"... .

Let us turn to ecology. I would like to recall the American writer Rachel Carson, who in 1962 published the book *The Silent Spring*. She essentially was the first to raise the alarm that life on the planet was threatened with annihilation. Scientists were also the first to warn about the destruction of the ozone layer and the greenhouse effect.

Let us recall the "great construction projects" of communism, which did such damage to the ecological balance. Did scientists really initiate them? Everything was done at the command of our "greatest architect." Scientists were the tools. Whoever resisted disappeared in the camps. Among all the scientists in the world there have always been both prophets, the first to warn of dangers, and careerists ready to do anything. And there were also forced laborers; this, unfortunately, applies to those Soviet scientists who in the past had to perform actions against their conscience.

If there is something to fear, it is anti-intellectualism. The public disgrace of Academician Sakharov, which "took place" in the Congress of June 2, made a strong impression on me. I think that only 300 pns of those present in the auditorium—independently of their relation to the position, statements, or style of behavior of Sakharov—imagine that many years from now history will preserve the name of this People's Deputy of the first Congress. It is truly terrible when all misfortunes are blamed on intellectuals, when intellectuals are spoken of with irony, when they are hounded, for example, by workers. Then twilight settles over the country... .

I have often had the possibility of observing the work of a Soviet embassy abroad. And I have often been asked to share my impressions of their work. In general, how does the activity of our entire foreign service establishment appear? Am I satisfied with the glasnost in this area? It is an interesting question.

Now everything is different here. In particular, I have established quite good relations with the Ministry of Foreign Affairs. In this entire hierarchy I feel sincere interest in having our scientists actively participate in shaping our foreign policy, in developing a "scientists' diplomacy" as one component of popular diplomacy. And we are obtaining practical assistance in all our affairs and undertakings. I say this as Chairman of the Soviet Pugwash Committee.

I received new confirmation of these impressions of mine when as a member of the Committee on International Affairs of the Supreme Soviet of the USSR and then in the Supreme Soviet itself I listened to the presentation of Eduard A. Shevardnadze and his responses to the numerous questions. It is not an accident that according to the results of popular opinion polls, the minister of foreign affairs has the highest rating (speaking in chess language) among his colleagues in the government: 82 out of 100 possible points.

Now about the embassies. Even in the past we had more or less normal relations with our foreign service. We felt the basic concern of the embassies, that is, the dilligent performance of the formal functions assigned to them regarding us, as citizens. And it is all the more gratifying that now our relations (I have in mind both simply delegations of scientists and of the members of the Soviet Pugwash Committee) with the foreign service have acquired qualitatively new content. Whereas earlier ritual (meetings, farewell parties) and, so to speak, the "performance of duties" dominated, now there is more lively interest in the activity which we have come to that country to take part in. Whereas earlier our embassy staff were interested in, roughly speaking, scientific-technological novelties, which we, they assumed, must know about, now they are interested in our ideas about applying the experience gained in other countries at home. The talk is now not of collecting information for a report or for depositing in the safe, but of how it can actually be used for the good of our country. What can be adapted for our land? What cannot? How can "their" specific features be taken into account? How can ours? In a word, the approach is much more serious and practical.

Accordingly, our new relation to the work of the embassy (and it is obvious) dictates its own requirements on the personality of the ambassador himself. I therefore consider it highly inappropriate to treat the embassy as a place of exile, where people who have compromised themselves in domestic politics are transported to the high post of ambassador of the Soviet Union. In my opinion, this practice, which still goes on today, must be ended.

In general, all our foreign service must be composed of high-class professionals. I am not speaking of a knowledge of foreign languages. And what a paradox we have here: it is considered necessary for all embassy workers to know the language of the country in which they serve, but sometimes an exception is for some reason made for the ambassador. Isn't this strange?

There is another problem. Now in many forums in the Ministry of Foreign Affairs and in some publications there is often talk of the acute need for the people involved in making decisions to have various options, a choice of alternatives. But how can this be ensured?

I think that we simply must have a large number of independent centers specializing in political science. In fact, if not only the U.S. establishment but also we attentively pursue the ideas produced by the RAND Corporation, then why shouldn't we have the same structural link in our foreign policy mechanism?

But we are speaking not of a single link, but of the guaranteed existence of an entire spectrum of viewpoints on a problem. This is clearly needed not only for "mind games;" it is the main condition for breaking the monopoly on truth. Is it necessary to state how harmful such a monopoly is?

We do not need to go far to find examples. Let us turn not even to political science, but to the field of expert opinions. Say we have the following problem: how and from what distance is it possible to detect by physical means the presence of nuclear warheads on missiles on board a ship? A great deal has been said about this problem, which is extremely important both for concluding the corresponding treaty, and for maintaining the balance of forces. Here in the USSR we have had many specialists working on this problem, and they have arrived at different conclusions. Some have maintained that nuclear ammunition can be detected on board a ship at a distance of several kilometers, while others disagree with this and show, with facts and figures, that at best the distance is 100–200 meters. Initially the first point of view was accepted, but then it was shown to be unsound. This is possible in physics and mathematics: no one will begin to seriously question the validity of, for example, the multiplication table. But what about in political science?

Here there may be, and in fact it is natural to have, several opinions about a particular situation and several sound evaluations. And this by itself should put one on guard and make one deal with the question more seriously. This means that analyses, reviews of the situation, discussions, and knowledge of various tactical measures for a particular strategic goal are necessary. In brief, the grounds for each decision must withstand serious opposition, the necessity of some action must be demonstrated to both its opponents and to society, and not made to "conform" to some anonymous department. Only in this way can we elaborate our politics, and not just correct the "sins" of the past.

The press plays an extremely important role here. Now the attention of the entire public is turned to it. Claims are made on it from different positions. For example, we ask: does our press reflect the entire spectrum of opinions, the full complexity of the internal political process? I don't think it does yet. But if we recall from where we began, the information available to us today seems a great gift.

In fact, if some historian had thought to write a treatise on the topic "The USSR in the 1970s" using exclusively newspapers, he would have gotten a more or less accurate picture of the country only from the western press. It would have been absurd to look for some "precrisis" in our publications of those years. It is not without reason after all that those years saw the birth of the witty saying "In *Pravda* ("Truth") there is no news, no "izvestiya," and in *Izvestiya* there is no truth"... .

And still now, in the era of perestroika, everything is not all right. For example, in 1987, after the publication in *Pravda* of an article in honor of the hundredth anniversary of the birth of Yan É. Rudzutak (who was shot in great purges, but *Pravda* kept silence on that point), I wrote to the editor of our leading newspaper a letter entitled "On the limited truth in *Pravda*." I received no answer. However, in the letters department I was told that they have a deadline of three months for answering. When this time had passed, I wrote a personal letter to the head editor Comrade Victor Afanas'ev, but it also went unanswered. It later became clear to me that in many questions regarding the history of our country and literature and art it was problematic to seek objectivity and truth in *Pravda*. Of course, as far as the popular diplomacy and activity of scientists in this area, and foreign politics in general, is concerned, most of our main newspapers are trying to do great and important work.

And in order for this work not to suffer, it is necessary to give our press legal guarantees. At present there are none; after all, our experience is well known: a seemingly good law is passed, and then it is reduced to nothing by various corrections and "clarifications." What about, for example, the unfortunately well known decree that journalists be given special access to certain events which has to be obtained from the Ministry of Internal Affairs? I think that this idea is akin to the limitation on subscriptions in 1989.

So we need a clearly stated law governing the press, which fixes both the rights of journalists and the rights of people who, for example, experience defamation. Here the courts must decide. The rights of the press itself are also important. I think that it should be free to discuss any question, to print any discussions and articles, except, of course, appeals to violent overthrow of order, and xenophobic and pornographic material. Ultimately, if the press operates according to the principle "what is not forbidden is allowed," life will become simpler not only for it but also for the government. At present, unfortunately, the versions of a law governing the press with which I have become acquainted do not inspire optimism. I think the two expressions "democratic perestroika" and "bureaucratic version of perestroika" have appeared for a real reason.

I am glad that there have been many practical and sensible speeches made at the Congress of People's Deputies of the USSR. What has been said by Shmelev, Popov, Afanas'ev, Bunich, Emel'yanov, the specific program which they have presented, all seemed to me an important contribution to resolving the eternal problem of "What should we do?" Of course, this is only a part of those ideas which have been discussed and argued over in sessions of the so-called "Moscow group." I intentionally say "so-called," because the session of the group which I attended had 300 people, among whom the Muscovites made up no more than 10%. Thus, this group contains not only Muscovites, but actually is simply a group of people with similar ideas forming an interregional deputies group.

As far as social and political problems are concerned, here there is an extremely sharp difference of opinions.

The polarization of forces in social and political problems seems to me to arise primarily from disagreements over the question of what should be given priority: universal human values or something else. It would be useful to pose the question: why have we had our greatest successes in our foreign politics? The problem seems to be that, having implemented the principles of the new political thinking in our foreign policy, and primarily the principles of the priority of universal human values, we are not in a hurry to put them into "domestic" practice.

USSR Foreign Policy in a New Dimension

Within the context of international relations entering a qualitatively new era, the foreign policy of the Soviet Union is distinguished by a single important feature: it is based on the absolute priority of universal human values over all others. There is no doubt that perestroika has brought tangible positive results in this area. It is very important that in recent years our foreign policy has acquired a number of features which were completely absent before, namely: the ability to listen to the arguments of the other side, even if they differ strongly from our own; the refusal to use force to persuade; a multifaceted search for mutually applicable solutions on disputed questions; a striving to achieve consensus and unanimity. These qualities are not something newly invented, but are necessary attributes of international relations in a civilized world.

It is important to note that the various documents brought before the Congress of People's Deputies and devoted to our foreign policy reflect discrepancies in their specific formulations. I suppose it is not really important which politician adheres to which views; each has his own ideas and philosophy. But it is worth discussing the most characteristic differences in defining the general nature of international relations.

> *We start from the class nature of international relations; any other statement of the question introduces only confusion into the consciousness of the Soviet people and our friends abroad.*

This is one position. Here is another:

> *Not excluding political and economic rivalry, peaceful coexistence is however not a form of class struggle.*

These are two rather clear concepts about which one can argue—a typical example of a pluralism of opinions.

The resolution of the Congress of People's Deputies "On the main directions in the domestic and foreign policy of the USSR" quite clearly describes our problems at the present stage, proposes the democratization of intergovernmental relations, and fixes the change in the course of USSR foreign policy. One can certainly speak of several different versions being put before the Congress, but the summary document clearly shows the position of the People's Deputies.

Speaking of foreign policy and the new role which the USSR Supreme Soviet has been called on to play here—its second session will open 25 September—it is impossible to ignore such an important aspect as glasnost in international affairs. Does this mean that we have to reveal all our plans in both the "minimum program" and the "maximum program"? Such extreme openness would, of course, run contrary to basic diplomatic methods, and even the entire centuries-old practice of international relations. It would be a mistake to assume that there do not exist any purely professional approaches, but, on the other hand, there should not be any secret diplomacy like that used in preparing the Molotov-Ribbentrop pact.

It would clearly be inappropriate to raise diplomatic dialog to the highest degree of openness and to broadcast the stage-by-stage preparation of international documents on television, but when it is the fate of the country that is in question, glasnost must be complete.

In connection with this, much is expected from the Committee on International Affairs. The Supreme Soviet and its agencies are supposed to do everything possible and necessary along the line of their interparliamentary relations so that the processes of perestroika, renewal, and glasnost become an integral part of the process of humanization of international relations in the interest of consolidating peace. In addition to studying specific foreign policy solutions and current and future problems and listening to the reports of Soviet governmental and parliamentary delegations sent abroad, we must consider the question which resounded dramatically in the Congress, namely, we must give a political evaluation of the past decision to send Soviet troops to Afghanistan, and after discussion by the Supreme Soviet we must report on the conclusions to the second Congress of People's Deputies. This is a real test of glasnost.

The Committee on International Affairs has quite close relations with the Ministry of Foreign Affairs. The Diplomatic Academy is ready to assist us in any way and provide information.

The Committee must pay special attention to questions about the history of international relations. This is extremely important because without knowing the past it is impossible to analyze the processes going on today. The dearth of information is especially acute in such areas as the history of how the

Second World War began, the situation in Greece in the late 1940s, the Korean War, the Berlin Blockade in 1948, and the events in Hungary in 1956 and in Czechoslovakia in 1968.

The other major tasks facing the Committee on International Affairs include preliminary discussion of drafts of laws and other documents concerning foreign policy which have been submitted for consideration by the Congress of People's Deputies and sessions of the Supreme Soviet, and the presentation of conclusions and proposals on a broad spectrum of international problems.

Perestroika in the sphere of international relations makes it possible to seriously challenge the philosophy based on confrontational thinking. With this in mind, the Committee will take control of the activity of the governmental institutions for performing obligations of our country, especially in the area of international economic collaboration, and will facilitate the development of relations with "people's diplomacy" organizations such as the Pugwash movement, Greenpeace, and so on.

Today various international agreements are acquiring great importance for us, since many of their statements are applicable to our domestic life. This is particularly true of the Vienna Summary Document: many times during meetings of our Committee questions arise concerning the correspondence between intrastate jurisdiction and the international legal system. So far in this area there are no direct contradictions. But if some arise, priority must be given to international legal norms.

Equally important are such things as the auxiliary protocol to the Geneva Conventions of 1949, which we subscribed to, concerning the protection of victims of armed conflicts not international in nature. The level of acuteness which our international conflicts reached owing to the distortions in our domestic policy is well known. We therefore must know which duties in this regard fall to the government and which powers are possessed by the agencies upholding order in light of existing international norms.

If broader rights in the economic arena are given to the republics, everyone will win. It is another issue when the question arises of leaving the Union. An equal sign should not be placed between the problems of economic independence and political autonomy; after all, history does not begin in 1939. It is first necessary to determine what will be the economic status of a small country if it obtains political independence. After all, it is no secret that even the republics in our country which are most well off economically cannot compete with the countries of the European Community. Will they have to exist as parasites, fed to stimulate their political views?

In my opinion, the foremost problem is to ensure the harmonic development of all the republics. The argument for this is: we have passed through joys and sorrows together; there is already a well-established political system, and the root cause of today's extraordinary tension in international relations must first of all be sought in our disastrous economic situation.

Russia's "Red-Brown" Hawks

Soviet nuclear specialists rail at vanishing funds and prestige; neo-fascist groups want arms and power—and a new Cold War.

Since the beginning of the "Gorbachev years," there has been a remarkable democratization in Russia, but there has also been a sharpening of internal contradictions. These years have led to a coalition of "red-brown" groups—the most reactionary elements of society—Stalinists, extremely aggressive members of the military-industrial complex, self-styled "protectors of the Russian Empire," national patriotic unions, and groups with their own storm troopers (who are, in fact, neo-Nazis). These have also been years of economic decline and the withering of research in the fundamental sciences.

In the early summer of 1990, when the conflict between Mikhail Gorbachev and Boris Yeltsin was revealed, it greatly revitalized the activities of military officials and nuclear scientists closely connected with the military-industrial complex. Some military circles had continued to think seriously about nuclear war even during the Gorbachev years. Just as a Bundeswehr investigating team found plans in the East German archives for a Warsaw Pact attack on Central Europe involving tactical nuclear weapons, similar plans were made even after the Warsaw Pact's May 29, 1987 "Berlin Declaration" that called for abandoning the "enemy image" and reliance on force. [1] (The Eastern strategists planned to reach the English Channel in two weeks, the Mediterranean Sea in four.)

Shortly before the August 1991 coup, Soviet nuclear scientists began to air their complaints in public. In an interview in the June 26, 1991 *Komsomolskaya Pravda*, Yevgeni Avrorin, the scientific director of Chelyabinsk-70, said: "In all likelihood, at the present time we are taking our first steps toward World War III. Peace can only be preserved by keeping superpower parity. Yet we [the Soviet Union] are disarming. Most initiatives made in the last years, such as a one-sided moratorium on nuclear tests and a reduction of the financing for defense sciences, are simply unilateral disarmament obscured by loud rhetoric."

Boris Litvinov told *Komsolskaya Pravda*: "Maybe we need to decrease our nuclear arsenals and tanks, the production of which has been excessive. Then, instead of hundreds of existing units, we scientists must design a single new warhead that will allow us to maintain parity. The government is not giving us such an opportunity."

And Professor Vladimir Nechay, director of the Research Institute of Technical Physics (Chelyabinsk-70), told the notorious *Den* ("The Day"): "Nuclear weapons are designed to destroy military objects first. It is not necessary to destroy New York or Moscow. Missile silos, military bases, and command posts will be hit first... . If it costs a hundred explosions... nothing terrible will happen. I think that our half-schooled ecologists have caused more harm to the people's health than all the nuclear tests carried out in the country together."

But on August 8, 1991 Academician Yuli Khariton, scientific director of the Research Institute of Experimental Physics (Arzamas-16)—and the remaining patriarch of Soviet nuclear weapons—told *Izvestia*: "On July 30 and August 2 the Leningrad TV station aired the program *Panopticum* ("600 Seconds"), which covered our defense industry, highlighting the classified research town of Arzamas 16, where I work. The authors of this program, Col. Nikolai Petrushenko, a Peoples' Deputy, and TV anchor Aleksander Nevzorov, whom I did not know before, dishonestly used my name for their own political purposes: to stir up mistrust toward the new START Treaty... and to attack the Soviet leadership's policy as tantamount to treason. I strongly object to such a position."

Khariton wrote of the necessity and unavoidability of cutting the military budget while international conditions were radically changing and the country was experiencing a deep economic crisis. But he also stressed the necessity of preserving such unique scientific and design collectives as Arzamas-16 and Chelyabinsk-70.

The identification of the U.S.-Soviet nuclear weapons reduction treaties as high treason, along with policy statements issued by the leaders of the August coup, were ominous. During the coup, a statement on the renewal of the Cold War was prepared by the Ministry of Foreign Affairs, but it was never issued. The coup leaders did issue another diplomatic document, one that informed Saddam Hussein of their desire to "restore" the Soviet Union to its position as a counterweight to U.S. influence in the Middle East. The junta's letter to Saddam Hussein was repudiated immediately after the legitimate government was restored.

The loss of nuclear science and technology's prestige can be illustrated through the example of the Moscow Institute of Engineering and Physics, whose mission was altered in 1943 to prepare specialists for the creation of Soviet nuclear weapons. Among the founders of this institute were Igor Kurchatov, Yakov Zeldovich, Isaak Kikoin, Lev Artsimovich, and other outstanding Soviet nuclear scientists. This institute belonged to the Atomic Ministry. Joining this very privileged institution was extremely difficult; not only were there stringent rules covering an applicant's biographical particulars, there were also up to 20 applicants for each vacancy. In recent years, applications have fallen to less than one per vacancy. In 1991, the best students selected by the capital of the unofficial "Nuclear Archipelago," Arzamas-16, refused to go there.

When they eliminated the American nuclear monopoly, Soviet nuclear arms specialists were cast as saviors of the world and saviors of the Socialist Motherland. Now, especially after Chernobyl, with the difference between "peaceful" and "military" atomic energy shrinking in the minds of millions, nuclear scientists, as well as chemists, are viewed as if they were poisoners of our planet, capable of killing nature—plants, animals, and even people—merely to satisfy their self-interest. Moreover, stories have been emerging in which KGB intelligence officers are trying by any means to disparage the scientific merits of Soviet scientists, and to persuade the public that the most important work on Soviet nuclear weapons was performed by the KGB officers who obtained information through espionage.

The early nuclear specialists also received many government awards—Hero Stars, Orders, Lenin and Government Prizes, high salaries, high-quality medical care, food, and goods. To some extent, all these rewards compensated for the impossibility of traveling abroad, for life behind barbed wire, and for the tough regime.

Today all the advantages of living and working in nuclear arms centers are gone, except for the availability of superb high-tech equipment and computers, an advantage that cannot be found in the "civil sciences."

In part to prevent the flight of Russian nuclear specialists, the U.S. Congress appropriated about $400 million for the dismantling of Soviet nuclear weapons. In addition, an initiative by three foreign ministers—Andrei Kozyrev of Russia, Hans-Dietrich Genscher of Germany, and James Baker of the United States—led to plans for a center in Moscow designed to open up new opportunities for Russian scientists, to prevent their emigration, and to become a laboratory for the development of new technologies.

Because hundreds of millions of dollars have been set aside for very specific purposes, the question of their proper distribution has arisen; many believe that the funds should be disbursed only to people who are directly involved in weapons dismantling, not given to those who work in related areas. According to some arms specialists, "the physicists from Dubna and the Kurchatov Institute know no more about nuclear weapons than is written in any textbook."

According to Pavel Felgenhower, the new Moscow science center will "help some Soviet scientists financially, but it won't prevent the proliferation of nuclear weapons. On the contrary, it will most likely stimulate the emigration of true military nuclear specialists." [2]

The problem of a "brain-drain" in the nuclear sciences was downplayed by Minister of Atomic Power Victor Mikhailov in November, 1991: "So far not a single qualified nuclear scientist or nuclear weapons designer knowing important information has left the country. With confidence, I can say that my colleagues are patriots and won't leave the country to design nuclear weapons for anyone." [3]

Let us hope that Mikhailov is right, and that patriotism will play a positive role in preventing the flight of nuclear specialists. But at the same time we should keep in mind that the so-called "national patriots," including members of the neo-Nazi society *Pamyat*, consider the United States and Israel to be Russia's main enemies and Saddam Hussein and Muammar Khadafi its best friends and closest allies. As *Nezavisimaya Gazeta* pointed out, *Pamyat* is very active in Arzamas-16. Most members are mathematicians, inspired by strongly anti-Semitic colleagues like Igor Shafarevich, but some are physicists. Some of them might be willing to design an "Arab bomb," especially if they were paid good salaries in hard currency.

In contrast to Russian denials of a "nuclear exodus," German Federal Intelligence (the "BND") claims to have extensive files on the recruitment of Russian and other Eastern experts by Third World countries. According to the BND, several scientists are reported to have signed an agreement with Algeria. Four others offered their services to India. Fifty specialists, including one from Arzamas-16 and a missile specialist from Ukraine, currently live in Iraq. Fourteen atomic specialists have been residing in Iran since 1991, and new work agreements with Teheran have been signed by 50 engineers and 200 technicians. Two Russian specialists received offers from Libya, but many Russians were already involved in a number of unofficial nuclear projects there. Last but not least, among the 7,000 scientists that have immigrated to Israel since 1989, 40 are atomic specialists. [4]

In order to stop the "nuclear exodus," a second international scientific center was to be created, to deal with employment problems within nuclear manufacturing. This venture was supposed to be supported by the European Community, but the center's construction appears to be moving slowly.

Within a year or two, the United States and Russia will sign an agreement on the purchase by the U.S. Energy Department of enriched uranium from dismantled Russian weapons.

The Americans tried to obtain from Russia all documents related to the export of uranium, not only to the United States, but to other countries, as well as documents on uranium enrichment. By Russian demand, the market price of uranium at any point in time will be announced by the U.S. Commerce Department only after consulting with the Russian Ministry of Atomic Energy. This Russian-American agreement is an example of agreements that stimulate the peaceful applications of atomic energy, encourage nuclear weapons dismantling, and prevent nuclear proliferation.

Other arrangements may be less benign. There are many rumors about the illegal export of nuclear weapons, especially from Kazakhstan to Iran. In March 1992, the BND said that Iran had purchased two nuclear warheads from Kazakhstan. [5]

Citing U.S. intelligence as its source, on March 15, 1992 the Russian TV program *Vesti* ("News") reported that three air-launched missiles with nuclear warheads had been stolen from Kazakhstan, and that two of these warheads had been purchased by Iran. This story was categorically denied by the head of CIS military forces, who said that no tactical weapons remained on Kazakh territory in March 1992. Kazakh leaders also denied the rumor. During an official visit to Israel, Sergey Tereshchenko, the Kazakh prime minister, announced that his country was not planning to sell the nuclear weapons it inherited from the Soviet Union. Nevertheless, rumors continue to spread about a Kazakh-Iranian accord. Kazakhstan is now releasing previously classified information on the size and characteristics of its uranium deposits.

Nuclear specialists from both sides have not yet laid aside their mutual mistrust. Yuri Trutnev, who gave Yeltsin the idea of retargeting Soviet missiles away from American cities, has bitterly noted that "American missiles are still pointed at our side." [6] Yevgeni Negin said, "Everyone is talking about disarming. But I am sure that the Americans don't even think about real disarming." [7]

At a January 14, 1992, meeting of the presidium of the Russian Academy of Sciences, Evgeni Avrorin suggested: "The question of whether Soviet nuclear weapons specialists run to Khadafi, Hussein, or any other aggressor is widely discussed in the Western press. To prevent this from happening, support for the work and the standard of living of scientists—not just tips—is needed."

It may seem ungrateful, but it is necessary to play the Cassandra role when forecasting Russia's future. The victory for democracy in August 1991 now seems to be irrevocably past. In enjoying their victory, the democrats failed to follow the advice of journalist Walter Laqueur: "This is no time to be generous to anti-democrats." [8] Instead of taking drastic measures against reactionary forces, the democrats began settling petty scores among themselves.

The situation in Moscow as well as the rest of the country is now characterized by massive demonstrations against economic disaster, hyperinflation, the failure of the government to keep its promises, and escalating crime, corruption, long queues—and high prices, which were previously unheard of. Armed groups of "national patriots" have been preparing to take action, and they recently attacked Moscow's most popular democratic newspaper, the *Moskovskiĭ Komsomolets.* The real civil war, with hundreds of thousands of victims, has been escalating in Tajikistan, the Northern Caucasus, between Georgia and Abkhazia, between Armenia and Azerbaijan, and between Moldova and Pridnestrovie. Cossacks are taking power in many areas, from the Don and Kuban to Priamurie.

Recently the Russian National Salvation Front, the most reactionary and fascist Russian organization, held a conference in one of Moscow's largest halls. The front is led by TV anchor Aleksander Nevzorov, a sort of Russian version of Patrick Buchanan; right-extremist newspaper editor Aleksander Prokhanov; and others, including KGB Gen. Alexander Sterligov. Defense Minister Gen. Pavel Grachev's statement in support of the government didn't slow the extremists, and a breakdown in the army, if it occurs, will only complicate the situation in the country.

There is likely to be a significant reorganization of the government, which may require making an unstable peace with the right-wing coalition. In the best outcome, power would pass to a coalition of small centrist groups and the much-weakened democratic reform movement led by Gavriil Popov, the former mayor of Moscow. The centrist groups are currently led by Arkady Volsky, the leader of Civic Union, Aleksander Rutskoi, the vice president, and Nikolai Travkin, a former Supreme Soviet member. This coalition should lead neither to significant changes in Russian foreign policy nor in economic relations with other countries, nor in deviations from the current policy on nuclear disarmament. Nevertheless, the military-industrial complex would play a more important role in the future—mainly, it is hoped, to solve the problems of conversion.

A completely different forecast should be made if the Russian National Salvation Front comes to power. The group's theoreticians believe in the idea of a global confrontation between the forces of Orthodoxy and Islam on the one side and Catholicism and Judaism on the other. The United States, Japan, and China would be considered Russia's strategic enemies. Nuclear disarmament would become unpatriotic, and the continuation of nuclear testing would be a vital task. The Russian National Salvation Front's foreign policy constitutes the real Cold War. The world community should remember this well, and base its future relations with Russia and CIS on that knowledge. Only a democratic Russia has the right to obtain political support and economic help from the Western democracies.

1. Vladimir Milyutenko, "Did Soviet Generals Dream about Bonn, Brussels, and Paris?" *Nezavisimaya Gazeta*, August 17, 1991.
2. Pavel Felgenhower, "Russian Closed Towns," *Nezavisimaya Gazeta*, June 30, 1992.
3. Victor Mikhailov, "Will our 'Brains' Flee?" *Komsomolskaya Pravda*, November 27, 1991.
4. Yevgeni Bovkun, "Nuclear Arms Specialists are Flying from the CIS," *Izvestia*, October 19, 1992.
5. Victor Litovkin, "Generals in Moscow Categorically Deny Reports on the Sale of Nuclear Weapons," *Izvestia*, March 16, 1992.
6. Yuri Trutnev, "A Bomb is the Cheapest Way to Avoid a Nuclear War," *Izvestia*, August 18, 1992.
7. Yevgeni Negin, "Nuclear Sunset," *Pravda*, April 16, 1992.
8. Walter Laqueur, "This is No Time to be Generous to Anti-Democrats," *International Herald Tribune*, September 19, 1991.

Scientists Excepted[1]

I must express my categorical disagreement with the headline you arbitrarily gave to my article ("Russia's 'Red-Brown' Hawks") in the June 1993 *Bulletin.* By no means would I apply the term "red-brown" to any scientists—including those experts in the field of nuclear weapons referred to in my article. Only certain notorious political figures mentioned in the article, well-known for their reactionary, Nazi-like activities, deserve the description "red-brown," which is widely applied to them in Russia.

Unfortunately, readers may have obtained the false impression that I applied the term "red-brown" to nuclear weapons experts.

Vitaliĭ I. Gol'danskiĭ
Semenov Institute

[1] Letter to the Editor of *The Bulletin of the Atomic Scientists.*

ANTI-SEMITISM, EMIGRATION, AND THE BRAIN DRAIN

The subject of this section is particularly painful for me—both because I am Jewish myself and because I was brought up and educated in the traditions of great Russian and world humanistic culture and in the spirit of internationalism.

The similarity and tight relationship between the Hitler and Stalin regimes caused an easy spread of pathological anti-Semitism in the Soviet Union. Governmental anti-Jewish policy (championed in the ill-famed "unmasking" of "doctor's plot" when only Stalin's death prevented the mass deportation of Jews to the Arctic) effectively promoted the revival of the old Russian nightmare—daily popular anti-Semitism, memorable since pogroms and "Black Hundred" outrages in Tsarist Russia.

There is certain mutual stimulation (positive feedback between the anti-Semitism of bureaucrats at all levels) of millions of ordinary people suffering from disastrous economical situation and the militant neo-Nazism of various growing "National-Patriotic" forces and movements. There are at present in Russia more than 150 legal anti-Semitic newspapers and journals, open selling of "Mein Kampf," "Protocols of Elders of Zion" etc., and the open wearing of swastikas. The situation has by no means improved since the time of publication of my article (No. 47) in the *Washington Post* and in *Physics Today* (February–March 1990). The response of notorious "Soviet Russia" to that article represents the "soft edge" of the broad spectrum of such responses of anti-Semitic forces, while the "hard edge" was represented by anonymous letters declaring the death sentence for me. Our top authorities still try to avoid the resolute condemnation of anti-Semitic activities, the irreconcilable and merciless struggle against the neonazists stormtroopers.

Paradox of Perestroika: Ethnic Turmoil and Anti-Semitism[a]

A specter is haunting Soviet society. The specter is the economic, cultural, and political turmoil that a writer in the government news paper *Izvestiya* likens to "a fever... that weakens *perestroika...* and deafens the leadership to the voices that cry out in alarm to prevent disorder and anarchy." As democracy and diversity sweep through much of Eastern Europe, the Soviet behemoth is mired in ethnic upheavals and nationalist upsurges. Such is the nature of the troubles in this era of *glasnost* that extremist groups with names like Fidelity, Renewal, Fatherland, and Memory—all xenophobic and ostensibly devoting themselves to the preservation of historic buildings, churches, and monuments, combating drunkenness or reviving Russian folksinging—are waging a rearguard action, deploring Western-style freedoms and advocating anti-Semitism. The actions of these groups cause Soviet Jews to think about a modern exodus.

Though worries about anti-Jewish outbursts have been increasing for nearly two years, anxieties were inflamed by recent events—notably the disruption of a meeting of "April," the liberal writers' society, at the Central House of Writers in Moscow on the evening of 18 January. A gang of about 30 bullyboys, members of a paramilitary group called *Pamyat*, broke into the hall, shouting, "Yids, get out to your Israel" and "Today we are here with loudspeakers; tomorrow we will come with guns."

Anti-Jewish episodes are becoming bolder. Jewish cemeteries have been desecrated. Since late January, posters have appeared on Moscow's Arbat Street with the text "If you are Russian, kill the Jewish." *Pamyat* has gone so far as issuing a blacklist on which appear the names of prominent Jewish

[a] By Irwin Goodwin.

physicists such as Arkady Migdal. In addition, leaflets are being distributed warning that pogroms will take place in May.

To many Soviet Jews the organized hatred is a paradox. Conditions for Jews have improved considerably under Mikhail S. Gorbachev's policies. Synagogues have been allowed to reopen. Jewish books and newspapers, though still few in number, are being published again. Jewish cultural societies have sprung up in Moscow, Leningrad, Kiev, and many provincial cities, especially in the Baltic republics. Jews are being admitted to more universities and taking command of the news media. Many more Jews are being allowed to emigrate, and more visas have been issued for Jews to travel.

Yet memories of tsarist oppression and Stalinist violence run deep. The mood of Soviet Jews hovers between pessimism and panic.

Soviet leaders have been curiously silent. Last year, a Lithuanian Jewish writer, Grigory Kanovich, joined by two members of the Soviet Academy of Sciences, physicist Vitaly Ginzburg and biophysicist Oleg Gazenko, submitted a petition to Gorbachev, signed by more than 200 of their fellow members of the Congress of People's Deputies, calling on him to condemn anti-Semitism and to create a special committee to investigate the problem. The petition was buried. A few weeks later, ten prominent scientists and writers sent a letter on the subject to Gorbachev (see below). No formal answer was received—though one of Gorbachev's associates informed them that their message was discussed with Politburo members.

Then, on 21 February, Soviet prosecutors indicated they had begun a criminal investigation of *Pamyat*, the most aggressive faction of the anti-Jewish extremists. The Interior Ministry announced the same day that "rumors of pogroms against Jews... have no grounds whatsoever." The statements did not allay fears. As one Jewish physicist told an American colleague, "Whenever a stranger comes to my door I think I am about to be shot dead."

Jewish scientists have long been prey to open and furtive discrimination. But after the end of the Stalin era their condition improved—though they could still be sent to prison or gulags for acts considered hostile to Soviet policies, such as participating in human rights demonstrations. Refuseniks were fired from their jobs for no other reason than that they applied for an exit visa.

Today's enemy of Soviet Jews seems to be, ironically, *glasnost* and *perestroika*, which have unleashed forces that appear out of control. Gennadi I. Gerasimov, spokesman for the Soviet Foreign Ministry, told an interviewer recently that "we have no history of democratic experience. Freedom to some means anarchy."

When the Soviet public contemplates the harsh realities of food shortages, urban crime, and ethnic strife, along with the deterioration of authority, they may wonder whether their leaders, Gorbachev included, aren't just fiddling while the USSR burns.

In his column in *The New York Times* on 12 February, William Safire wrote that "with their inborn barometers of bigotry, Soviet Jews sense a coming disintegration of authority and know they will be scapegoats in any return to ethnic passions." While Soviet authorities are unsure of the exact number of Jews in the country, U.S. groups estimate that about 1.5 million consider themselves Jews. Increasing numbers seek to emigrate.

Indeed, Soviet scientists are coming to the U.S. as immigrants and visitors in record numbers. "It would be easy for any one of my colleagues to get a tenured professorship here," says a visiting Soviet physicist. "But I don't know any who want to emigrate. Still, the Jewish ones are frightened about the political instability and the anti-Semitic insanity."

In the past year or two, "Little Russias" have emerged at such universities as MIT, Princeton and the University of Minnesota. Yuri Cherniak, a planetary physicist at MIT who once ran the refusenik science seminar in Moscow, insists that Soviet authorities aren't likely to improve conditions for Jews—and that negligence, he says, will impel more scientists to "join the brain drain" from the USSR.

One who understands the current situation is Vitaliĭ I. Gol'danskiĭ, director of the Semenov Institute of Chemical Physics and a member of the Soviet parliament. Gol'danskiĭ recently wrote a pointed and poignant essay; it is published in this issue, beginning on page 53.[1]

[1]See next chapter.

On "Special Dangers" of Perestroika to Soviet Jews, Science, and Society

Supporters of Mikhail S. Gorbachev's *perestroika* are increasingly alarmed by the possibility that this program of economic restructuring and reforms may collapse. Should this occur—and it cannot be ruled out, even in the near future—it would be a disaster not only for the Soviet Union but for all humankind.

Many of the difficulties already encountered by *perestroika* are well known outside the Soviet Union, as are some of the potential consequences if the program fails. But too little attention has been given, until recently, to the special dangers for Soviet society posed by the increasing aggressiveness of extremist right-wing[1], virulently anti-Semitic groups that seek to subvert *perestroika*, to blame the country's past and present problems on the Jews, and (as some of their propaganda states explicitly) to "finish what Hitler started."

These extremists are flourishing in a climate of spite, envy, scapegoating, and hatred associated with the increasingly severe troubles in the Soviet economy and the steadily mounting tensions in ethnic regions. They are perhaps already the strongest, and certainly the fastest growing, of the divisive forces pushing the country toward bloodshed and, in the worst scenario, civil war.

The extremist groups go by a variety of innocuous-sounding names, of which the "National Patriotic Front *Pamyat*" (*pamyat* means "memory") is the best known outside the Soviet Union. A number of them recently entered

[1]See author's note on page 139.

into a confederation under the title of "Bloc of Social-Patriotic Movements of Russia." I prefer to call them Russian monarcho-Nazis (or monarcho-fascists), to reflect the combination of their deep reverence for the autocratic tsarist Russian empire and their ferocious hatred of Jews.

Incredibly, these monarcho-Nazis openly condemn the Jews as the main (if not the sole) culprits in all of Russia's troubles from the 1917 October Revolution up to the present—including genocide against the Russian people in the form of the millions of deaths during the civil war, collectivization, and various purges; destruction of tens of thousands of Russian churches and historical monuments; and spiritual poisoning of the people through the introduction of "decadent" and "corrupt" Western culture alien to Russian tradition. These extremists also accuse the Jews of ritual murders and of a worldwide conspiracy against humankind, making reference to the falsified and discredited "Protocols of the Elders of Zion."

There is a striking similarity, in fact, between the views, programs, and intentions of the Russian monarcho-Nazis and the original Nazi platform as laid out in Hitler's *Mein Kampf* and other infamous documents of the period. The resemblance of the current situation in the Soviet Union to that of Germany in 1931–33 has been publicized by progressive Soviet mass media in recent weeks.

The principal organization serving as a coordinator of the monarcho-Nazi forces is the Union of Writers of the Russian Federation (RSFSR). As outlets for their propaganda the extremist groups have at their disposal such newspapers and journals as *Literaturnaya Rossiya* (Literary Russia), *Nash Sovremennik* (Our Contemporary), *Molodaya Guardija* (Young Guards), and *Moskva*. The leaders of this movement include many writers, some scientists, some artists, and others.

The speeches and publications of these Russian neo-Nazi groups are becoming routine features of everyday life in the Soviet Union. Professor Herman Andreyev of Mainz University in the Federal Republic of Germany analyzed their form and content in a recent issue of the weekly magazine *Ogonyok*. He concludes that in Western European countries such statements would be treated as unconstitutional, the persons propagating them would be called to account, and the organizations supporting them would be dissolved.

Yet the monarcho-Nazis seem to be meeting no serious opposition—indeed, more often sympathy and connivance—from important party and government leaders of the USSR. It is instructive, for example, that the platform of the Soviet Communist party on ethnic problems, published in August 1989, said not a single word about the anti-Semitic campaign against so-called "cosmopolites" (1949), the shooting of leading Jewish writers and artists (1952), or the disgraceful episode of the "doctors' plot" (1953), when it was alleged that several Jewish physicians and surgeons were attempting to murder Stalin. By contrast, many other Stalin-era crimes against various nationalities of the Soviet people were scrupulously singled out in the party platform.

Similarly, in June 1989 more than 200 people's deputies expressed their concerns to the Presidium of the First Session of the Congress of People's Deputies about the "growing wave of anti-Semitic activities, including open calls for violence that could lead to irremediable consequences." The appeal went unanswered. That was also the fate of a letter on the subject that ten distinguished scientists and writers sent to Gorbachev the following September (see page 54).[1]

The explanation of such passivity on the part of the authorities seems quite simple: It is the evident sympathy of many Soviet government officials on different levels to the views of the monarcho-Nazis. Others in positions of authority who do not sympathize nonetheless hesitate to act because of the way the growing aggressiveness of the extremist groups is linked to the bloody ethnic conflicts and intensifying separatist movements in nearly all of the outlying districts of the USSR.

Specifically, this situation offers the monarcho-Nazis considerable opportunities to blackmail and intimidate Gorbachev and his closest advisers by claiming that with the "decline of the empire" the Russian heartland and her "genuine sons" constitute the only reliable saviors of Gorbachev's power. Such arguments are being used to push Gorbachev toward the right and to divide him from his true supporters on the left—the liberal intelligentsia. The result could be a repetition of the circumstances that led to the downfall of Nikita Khrushchev in 1964.

In parallel with their attempt to intimidate Gorbachev, the monarcho-Nazis have been openly attacking his foreign as well as domestic policies and the underlying principles of his "new thinking"—the primacy of universal human values, disarmament, ending the Cold War, and building a "common European home." They even have accused Gorbachev of being an agent in the service of the CIA and the Israeli intelligence service Mossad. With this two-pronged strategy of intimidation and direct attack, the Russian neo-Nazis hope to attain either a decisive influence over Gorbachev's policies or his removal from power and his replacement by members of their own movement.

What would that mean for Soviet Jews? The answer is all too clear from the similarity of the Russian monarcho-Nazi program to that of Hitler. The Russian extremists already possess their equivalent of the storm troops of the

[1]See next chapter.

SA and SS, in the form of the *Pamyat* movement. *Pamyat* does not disguise its intentions to carry out pogroms against the Jews, to whom it usually refers by the denigrating term *zhidy*, or "yids." In fact, members of *Pamyat* have been organizing well-attended meetings all over the country to call for pogroms—one such meeting even took place in Moscow's Red Square last 12 November—and no one has stood in their way.

Hitler treated as Jews those who have at least one Jewish grandparent. *Pamyat* goes further. It has announced its intention to search for Jewish progenitors back to the tenth generation. New recruits to *Pamyat* are required to prove their "racial purity" and to provide to the organization the home addresses of five Jews—no doubt to locate them for the pogroms to come. Opponents of the monarcho-Nazi movement who happen to be "racially pure," or "Aryan," are characterized, along with all the liberal intelligentsia, as "masons" or "*zhido*-masons" (that is, supporters of Jews). These are also the targets of pogrom propaganda.

The brazenness of monarcho-Nazi threats and actions against Soviet Jewry has been increasing. In addition to the anti-Semitic rallies and the desecration of Jewish cemeteries around the country, which have been going on for some time, it now seems that meetings of liberal intellectuals are no longer safe from disruption by *Pamyat* thugs.

On the evening of 18 January, for instance, a meeting of the progressive "April" group of writers at the Central House of Writers in Moscow was invaded by several dozen *Pamyat* thugs with megaphones. They roughed up some of the writers, forcibly ejected others from the hall, shouted anti-Jewish slogans, and announced that on their next visit they will be armed with automatic weapons. They also designated St. George's Day, 2 May, as the date for a pogrom. St. George, the dragon slayer, is a patron saint of old Russia.

Further increases in anti-Semitic activities (especially, of course, actual violence) surely will lead to a mass exodus of Jews, people of partly Jewish extraction and "racially pure" members of the liberal intelligentsia. This new wave of emigrants—refugees from monarcho-Nazi extremism—could reach several millions and would represent a serious brain drain from the USSR.

As for the possibility of another holocaust, it certainly could not reach the scale of the Nazi crimes. The world has changed too drastically in the last half-century for that to happen. But a wave of pogroms more or less along the lines of the infamous *Kristallnacht* cannot be ruled out—weaker if a government like the present one tries to oppose the neo-Nazi bigots, stronger under a government with monarcho-Nazi sympathies and a lust for pogroms.

What should be done? For starters, the worldwide public should be informed of the intentions and activities of the new followers of Hitler in the Soviet Union and should be told their names. The famous Brown Book published by anti-Fascists in 1933, after all, was the first important step in the exposure of the Nazi crimes of that era. Clearly, the publishers and editors of newspapers, journals, and books, as well as producers and writers in the electronic media have an important role to play.

The stakes are high. If the monarcho-Nazis prevail and *perestroika* collapses in an orgy of chauvinism and racism, the results are likely to include not only a rapid heating of an atmosphere of anarchy in the Soviet Union but even the possibility of a civil war. In a country bearing a huge stockpile of nuclear and chemical weapons, as well as possessing a far-flung system of nuclear power plants, a not implausible chain of events could quickly lead to not just a national but to an international catastrophe.

Letter to the General Secretary

Letter to the General Secretary of the Central Committee of the Communist Party of the Soviet Union, Chairman of the Presidium of the Supreme Soviet of the USSR, Comrade M. S. Gorbachev:

Dear Mikhail Sergeevich,

Basing our remarks on events in our country and on discussions held between June and August of this year with a number of prominent foreign political figures, scholars, and writers, we want to express in this letter (and, if possible, in person) our most serious apprehensions about what might happen in the next months and years if urgent measures are not taken to guarantee Jews safe residence in the USSR, unpoisoned by streams of ethnic insults, destruction of graveyards and direct threats of pogroms.

... in earlier times, the main source of emigration was those Jews who ... could be considered part of the Jewish culture because of their use of the ancient Hebrew language or of the Yiddish language that developed much later in Europe. Now, as a consequence of the increasing aggressiveness by several monarcho-Fascist groups, through acts and publications, Jews whose flesh and blood have been infused with Russian culture since childhood, whose spoken and written language is Russian and who make contributions, commensurate with their abilities, to the life of our country and to the development of our culture and science are being literally forced to flee our borders. It should be noted that as many as 75% of the Jewish population of the USSR may consider themselves in peril from actions taken or proposed by the extremist groups. In addition, it should be kept in mind that our monarcho-Fascists consider as Jews (or, as they call them, Judo-Masons) those who are half Jewish ... or one-quarter Jewish, those who have even a small amount of Jewish blood, and those who are married to or were married to Jews. (In the future, no doubt, these monarcho-Fascists are likely to search for signs of "racial inferiority" in each other.) In a word, the extremists, both in their propaganda and practice, are guided by the Nuremburg laws of Fascist Germany—right down to their demand for the prohibition of mixed marriages.

They compare the Jewish people to a parasitic vine that twists around a healthy tree and eventually drains the vitality of the tree. Their publications contain anti-Jewish characterizations such as "This is a people without a homeland." ... In their diatribes they assert that Jews exclude from their thoughts, speech, and writings such concepts as motherland, patriotism, *et cetera*.

Even Lenin's famous decree "On the Struggle Against Anti-Semitism" provokes a frenzy in members of the extremist groups. "If you open your mouth against the Jews, they will drag you around the corner and shoot you, and this will be covered up by the thunderous music of the 'Internationale,'" said *Molodaya Gvardiia* in a recent issue. An Italian woman, a teacher of history who accidently found herself at a *Pamyat* meeting in Leningrad, was moved to write, "It seemed to me that I was attending a replay of a Fascist rally during the 1930s." Demonstrations by *Pamyat* in Moscow, on Pushkin Square, are replete with signs, posters, and T-shirts that bear such slogans as " Down with the occupation of Russia by Jewish Nazis!" and "Down with the domination of the 0.69%!"

Every day, *Pamyat* agitators gather outside your offices in Moscow. Is it possible to imagine that in the center of Moscow they could go unpunished for insulting the national dignity of, say, Yakuts or Belorussians or demanding the exile of Ukrainians or Georgians?

Appeals such as ours in this letter, written in a multi-national state, acquire an explosive force. The work of the monarcho-Fascists is leading to the point where, in the coming years, the country may be deprived of 5–6 million of its best and brightest scholars, engineers, doctors, artists, and others. History has already taught the awful lesson of Nazi Germany and her Fascist allies. It is easy to see and be convinced which countries lost and which gained from the banishment of "racially impure elements" (like the Jewish physicist Einstein or the Italian physicist Fermi, who was married to a Jewish woman) and which people benefited or did not after Jews were forced into exile from their native lands ... .

It may seem to you that, against the background of the events in Nagorno-Karabakh, in the Baltic regions, and in Moldavia, the problem of anti-Semitism does not look so alarming or so primary in importance. As a matter of fact, events are occurring with such rapidity that, if they are not stopped now, the consequences may be irreversible. The situation has caused deep distress for many people in the Communist Party who remember the unlawful acts that were tolerated during the years of Stalinism and the miserable manifestations of anti-Semitism that were propagated from above, such as the campaign against "stateless cosmopolitans," the devastation and extermination of the Jewish Anti-Fascist Committee, and the notorious "doctors' plot."

In our view, your agenda at the forthcoming Plenum of the Central Committee of the Communist Party of the Soviet Union must include discussions of the most acute national problems. Your views, coming from the author of *perestroika*, about anti-Semitism in the USSR and its potential to destroy the chances for transforming our society must be expressed explicitly and unambiguously.

Dear Mikhail Sergeevich! *Perestroika* meets everywhere with the most fervent, impassioned sympathy. People want to campaign for *perestroika*; they want to promote it by all available means. Each of us, using the means available to us, has discussed possible ways of bringing this about with political figures, industrialists, and scholars here and abroad, and in these conversations many different projects and prospects became clear.

In the declaration adopted by highly authoritative scholars from 42 countries after the 39th Pugwash conference in Cambridge, Massachusetts, from 23 to 29 July 1989, an appeal was made to the world scientific fellowship—not without difficulty, we observe—to support *perestroika*. As a rule, Pugwash declarations do not contain statements dealing with internal matters of individual countries, but in the case of last July's meeting the conviction prevailed that the fate of our *perestroika* is likely to determine the future not only of the USSR but also of the entire world.

In such a situation, any shadow cast over our *perestroika* may lead to horrifying results, not only for the internal life of our country, but for world opinion toward the Soviet Union. The nationalities question is one of the indicators [of world public opinion], one of the most sensitive sore points. It is ironic that during the observance of the 50th anniversary of the start of the Second World War, the world feels especially sharply the danger of any kind of recurrence of Hitlerism and Stalinism, and the intelligentsia of the West very guardedly takes note that precisely in the years of *perestroika* the Soviet Union has become the focus for the most aggressive activities of neo-Nazis and monarcho-Fascists in Europe and for the most repulsive and massive propaganda on anti-Semitism.

Such are the external political impressions that we would like you to consider as part of our appeal for the active measures that we called for in the first part of this letter.

In conclusion, we want to repeat our request for a personal meeting.

Yours very respectfully,

Gregoriy Baklanov [editor in chief of *Znamya*, a literary magazine]

Yevgeny Yevtushenko [poet and People's Deputy in the Soviet parliament]

Andrei Voznesenskiĭ [poet]

Roald Z. Sagdeev [former director of the Space Research Institute and now head of the institute's theory division; a People's Deputy]

Vitaliĭ I. Gol'danskiĭ [director of the Semenov Institute of Chemical Physics and a People's Deputy]

Vyacheslav Ivanov [director of the State Library of Foreign Literature, a noted linguist and a People's Deputy]

Bulat Okudzhava [poet and popular folksinger]

Anatolii Rybakov [writer and author of *Children of the Arbat*]

Yuri Chernichenko [Soviet TV commentator on agricultural matters, essayist, and a People's Deputy]

Oleg Bogomolov [director of the Institute of Economics of Socialist Countries and a People's Deputy]

5 September 1989

Acknowledgments

The publisher and the author wish to express their grateful acknowledgment to the following publications and publishers for permission to reprint previously published articles. In each case, the material was re-worked for this publication.

I. Essays

A Living Legend: Lev Andrejevich Artsimovich is translated and adapted from *Recollections of Academician Lev Andrejevich Artsimovich*, Moscow: Nauka, 1981, pp. 106–112.

Aleksander Naumovich Frumkin is translated from *Aleksander Naumovich Frumkin. Ocherki, vospominaniya,...*, Moscow: Nauka, 1989, pp. 176–183.

On the 70th Birthday of Vitaly Lazarevich Ginzburg was not published before.

In the Kaleidoscope of Memory: Lev Davidovich Landau is translated from *Vospominaniya o L. D. Landau*, Moscow: Nauka, 1988, pp. 94–101.

Aleksander Nikolaevich Nesmeyanov is translated from *A. N. Nesmeyanov*, Moscow: Nauka, 1988, pp. 207–210.

Scientist, Thinker, Humanist: Andrei Dmitrievich Sakharov was originally published in *Physics Today* **43** (August 1990), pp. 47–49. It was translated from the Russian by David H. Parsons.

Fragments of the Past: Nikolaĭ Nikolaevich Semenov is translated and adapted from *Recollections of Academician Nikolaĭ Nikolaevich Semenov*, edited by Aleksander E. Shilov, Moscow: Nauka, 1993, pp. 43–62.

My Remarkable Relative Shura: Aleksander Iosifovich Shal'nikov is translated and adapted from *Aleksander Iosifovich Shal'nikov. Essays, Reminiscences, Miscellaneous Writings*, edited by Andrei S. Borovik-Romanov, St. Petersburg: Nauka, 1992, pp. 28–37.

Some Pages from the Past: Vladimir Iosifovich Veksler is translated and adapted from *Recollections of V. I. Veksler*, Moscow: Nauka, 1987, pp. 184–186.

Forty-five Years: Yakov Borisovich Zel'dovich is translated and adapted from *Familiar Unfamiliar Zel'dovich*, edited by Semyon S. Gershtein and Rashid A. Syunyaev, Moscow: Nauka, 1993, pp. 44–61.

What Should a Scientific Journal Be? is translated from *Izvestiya*, January 8, 1968.

Not Only about Chemistry is translated from *Znanie-Sila*, November 1984, pp. 7–9.

I Trust in the Human Mind is translated from *Yunost'*, No. 2, 1985, p. 97.

The Art of Developing Science is translated from *Sovetskaya kul'tura*, May 28, 1988.

Struggles of a Soviet Scientist is reprinted by permission of *The Bulletin of the Atomic Scientists*, **44** (8) (October 1988), pp. 33–35, © 1988 by the Educational Foundation for Nuclear Science, 6042 South Kimbark Avenue, Chicago, IL 60637, U.S.A.

Why Are We Lagging Behind? Remarks about Fundamental Science is translated from *Pravda*, April 5, 1989.

Science Has Had Its Say is translated from *Nauka I religiya*, No. 6, 1989, p. 12.

Let Common Sense Triumph is translated from *Izvestiya*, January 1, 1990.

Soviet Scientists Face Perestroika's Challenge is reprinted from *The Scientist*, **4** (4), February 19, 1990, pp. 15, 17. Copyright 1990, *The Scientist*. All rights reserved. Reprinted by permission.

Are the Discoveries of Lobachevsky Paying Off? is translated from *Moskovskaya Pravda*, April 15, 1990.

ON THE BASIC PRINCIPLES OF THE ACTIVITIES OF THE ACADEMICAL INSTITUTES is translated from *Vestnik Akademii Nauk SSSR*, No. 7, 1990, p. 127.

USING BOTH EYES is translated from *Moskovskii komsomolets*, January 5, 1978.

DEPTHS OF LIFE REFLECTED ON THE SCREEN is translated from *Sovetskaya kul'tura*, May 2, 1984.

HE ENTERED OUR LIVES: PUSHKIN is translated from *Nauka i zhizn'*, No. 2, February 1978, pp. 26–27.

ON THE ROAD TO RECOVERY is translated from *Sovetskii ekran*, No. 24, December 1987, p. 10.

SPIRITUALIZING THE MIND is translated from *Sovetskaya muzyka*, No. 4, April 1988, pp. 2–6.

DISASTROUS SHIFT TO THE RIGHT is translated from *Inzhenernaya Gazeta*, February 13, 1991.

RATING TV is translated from *Nedelya*, No. 25, June 1992, p. 7.

EFFECT OF A NATURAL CONTRACTION OF TIME was not published before.

II. Speaking Out

OUT OF THE SDI IMPASSE is translated from *Izvestiya*, March 21, 1988.

LOGIC OF THE MIND VS. THE LOGIC OF THE EGO is translated from *Izvestiya*, February 10, 1989.

ELIMINATION OF NUCLEAR ARSENALS was originally published in *Verification: Monitoring Disarmament*, edited by F. Calogero, Marvin L. Goldberger, and Sergei P. Kapitza, Boulder, Colorado: Westview Press, 1990, pp. 205–223.

AN INTERNATIONAL NUCLEAR SECURITY FORCE (approx. 19 pages) is reprinted from *A Nuclear-Weapon-Free World: Desirable? Feasible?* edited by J. Rotblat, J. Steinberger, and B. Udgaonkar (F. Blackaby, Executive Editor), © 1993 by Westview Press. Reprinted by permission of Westview Press, pp. 181–190.

COULD OUR COUNTRIES BECOME ALLIES? is reprinted from *Soviet Life*, August 1990, pp. 6–7, and was originally published in *Izvestia*, June 16, 1990.

UN NUCLEAR FORCE? A SOVIET SCIENTIST'S PROPOSALS is reprinted from *Arms Control & Disarmament*, FBIS-SOV-90-218, November 9, 1990, pp. 1–2, and was originally published in *Izvestiya*, November 9, 1990.

THE WORLD WANTS CLARITY AND RELIABILITY is translated from *Izvestiya*, January 10, 1992.

STATEMENT OF RUSSIAN SCIENTISTS was first published in *Izvestiya* and *Nezavisimaya Gaseta* on July 12, 1994.

AN ACADEMICIAN DEMANDS PUBLICATION OF THE MILITARY BUDGET is translated from *Le Matin*, December 3, 1987.

THE WORLD HAS BECOME EXTREMELY SMALL is translated from *Argumenty I Fakty*, No. 9 (438), March 4–10, 1989, pp. 6–7.

POLITICS, SCIENCE, AND GLASNOST is translated from *Mezhdunarodyaya zhizn'*, No. 8, 1989, pp. 122–128.

USSR FOREIGN POLICY IN A NEW DIMENSION is translated from *Za Rubezhom*, No. 39 (1524), September 22–28, 1989, p. 1.

RUSSIA'S "RED-BROWN" HAWKS is reprinted by permission of *The Bulletin of the Atomic Scientists*, **49** (5) (June 1993), pp. 24–27, © 1993 by the Educational Foundation for Nuclear Science, 6042 South Kimbark Avenue, Chicago, IL 60637, U.S.A.

SCIENTISTS EXCEPTED is reprinted by permission of *The Bulletin of the Atomic Scientists*, **49** (8) (October 1993), p. 54, © 1993 by the Educational Foundation for Nuclear Science, 6042 South Kimbark Avenue, Chicago, IL 60637, U.S.A.

PARADOX OF PERESTROIKA: ETHNIC TURMOIL AND ANTI-SEMITISM is reprinted from *Physics Today* **43**, March 1990, p. 52.

ON "SPECIAL DANGERS" OF PERESTROIKA TO SOVIET JEWS, SCIENCE, AND SOCIETY is reprinted from *Physics Today* **43**, March 1990, pp. 53–55.

LETTER TO THE GENERAL SECRETARY is reprinted from *Physics Today* **43**, March 1990, pp. 54–55.

Subject Index

About the Author

Born June 18, 1923, in Vitebsk, Belarus, part of the former Soviet Union, Vitaliĭ I. Gol'danskiĭ has lived through many hardships, yet has risen above the circumstances to achieve unparalleled success in not one, but several scientific disciplines: chemistry, physics, and biology.

The son of a Jewish teacher, Gol'danskiĭ was instilled with the ideals of communism at a very early age. He continued to believe in communism for several decades, experiencing bitter disappointments, which led him to become an active supporter of Khrushchev's thaw and Gorbachev's perestroika.

Surprisingly, Gol'danskiĭ's first passion was history (from age 10 to 15). In 1939, however, he began to study chemistry at the university in Leningrad. When war broke out, Gol'danskiĭ enrolled in a student construction battalion. While digging trenches he was wounded by a Luftwaffe bullet in 1941. Later, he suffered through a long, harsh winter during the siege of Leningrad and almost succumbed to death by starvation. Fortunately, he was evacuated in 1942. Shortly thereafter, he resumed his studies in chemistry at the university in Kazan' (at Volga River), where he also became gainfully employed as a laboratory assistant for the Academy of Sciences. This was the beginning of an illustrious career that spanned decades, culminating in the directorship of the Institute of Chemical Physics (ICP) of the Academy of Sciences.

In 1944, Gol'danskiĭ received a diploma in chemistry from the University of Moscow. He then began postgraduate studies at the ICP under Nikolaĭ Semenov—who would receive the Nobel prize in 1956 for chemistry—and Semenov's disciple, Nikolaĭ Chirkov. Research on hydrocarbon conversions and exploration of heterogeneous catalysis led to a dissertation for the Candidate of Chemical Science degree, which was awarded to Gol'dandskiĭ in 1947.

With the advent of the atomic bomb, the focus of many scientists and institutions shifted to nuclear physics. Gol'danskiĭ now turned his attention to experiments at the Dubna synchrocyclotron, involving the absorption and multiplication of high-energy neutrons. This work created the basis for Gol'danskiĭ's doctoral degree, the thesis (at that time classified) he successfully defended in 1952 at the Institute of Atomic Energy. Experimental results described in Gol'danskiĭ's dissertation and his subsequent reports and publications opened the door to what is now known as "electronuclear breeding"—effective production of plutonium (and its destruction, when necessary) by high-energy, high-current accelerators.

From 1952 to 1961 Gol'danskiĭ worked under Vladimir Veksler at the P. N. Lebedev Physical Institute of the Academy of Sciences (FIAN). In 1954 he published his first article in the Soviet Journal of Experimental and Theoretical Physics (JETP). Its subject was Čerenkov radiation from cosmic particles in the atmosphere.

Gol'danskiĭ and coworkers performed significant experiments on elastic γ-ray–proton scattering, leading to the determination of the electric polarizability of protons. Numerous experiments and theoretical considerations conducted later by various researchers did not change the value of this important constant obtained by Gol'danskiĭ's group. Another field of interest for the group was the high-precision studies on π°-meson photoproduction in hydrogen near the threshold.

Notable success was achieved by Gol'danskiĭ in describing the properties of nuclei far from the β-stability region and predicting the existence and characteristics of many new neutron-deficient isotopes that were then unknown.

Other predictions of nuclear phenomena made by Gol'danskiĭ that became milestones in physics included two-proton radioactivity, beta-delayed emission of proton and neutron pairs, and the nuclear Josephson effect.

In 1959 Gol'danskiĭ returned to his roots at ICP and delved back into chemistry experiments. Spurred by the discovery of the Mössbauer effect, he was now in a unique position, able to study chemical applications of this physical effect. Integrating chemical physics, nuclear physics, and radiochemistry, Gol'danskiĭ has applied the notion of nuclear chemistry to just such a combination.

More pioneering work led Gol'danskiĭ too predict and later to discover experimentally the quantum, low-temperature limit for the chemical reaction rate via molecular tunneling, i.e., temperature-independent, non-vanishing chemical reactivity even in the vicinity of absolute zero. For this achievement, Gol'danskiĭ was awarded the Mendeleev Gold Medal in 1975. The concept of quantum-mechanical tunneling of heavy particles in chemical conversions (particulary at low and ultralow temperatures) pioneered by Gol'danskiĭ plays a crucial and ever-growing role in contemporary chemical kinetics. It also contributes to the cold, extraterrestrial scenarios of prebiotic evolution and origin of life.

Gol'danskiĭ has been active in other scientific endeavors, too numerous to mention them all here, which include dynamics of biopolymers, physical chemistry of positron and positronium, and various fields of high-energy chemistry (e.g., radiation and shock-wave–induced chemical conversions in solids). He also found the time to teach these varied subjects.

Showing no inclination to curtail pursuit of his interests, this septuagenarian is now studying problems of the origin of life and of evolution on the basis of their relation to the spontaneous breaking of mirror symmetry in Nature.

As varied as are his interests, Gol'danskiĭ has received countless honorary memberships (e.g., the National Academy of Sciences, U.S.), numerous awards (e.g., the Alexander von Humboldt Award, Germany; the Lenin Prize, Russia), and lectureships.

His expertise and breadth of knowledge have afforded him the opportunity to serve as editor and editorial board member for several physics and chemical journals, to chair the Commission on Biological Physics of the International Union of Pure and Applied Physics, and to provide leadership on dissemination of Russian science and technology.

In view of the high esteem from his peers and the honors bestowed upon Vitaliĭ Iosifovich Gol'dandskiĭ, one may say that he is a man of many deep passions—chemistry, physics, and biology, as well as politics and pedagogy, to name a few—who is unique in the sense that he has pursued his interests to their fullest extent and expanded man's knowledge in so many fields.